W9-AIC-721

AN EXTRAORDINARY

Power
to Heal

by the authors of
FEAST WITHOUT YEAST :4 Stages to
Better Health
and
EXTRAORDINARY FOODS
FOR THE EVERYDAY KITCHEN

BRUCE SEMON, M.D., Ph.D.
and LORI KORNBLUM

Wisconsin Institute of Nutrition, LLP
Milwaukee, Wisconsin
www.nutritioninstitute.com

DISCLAIMER

An Extraordinary Power to Heal describes relationships that have been observed between the common yeast, *Candida Albicans*, certain foods, and medical conditions. It is not intended as medical advice specific to any particular person. Its intention is solely informational and educational for the public and the medical profession. Treatment of health disorders, including those which appear to be yeast connected, must be supervised by a licensed health care professional. Either you or the health care professional who examines and treats you must take responsibility for uses made of this book. This book is not intended to create legal advice for any particular person, nor is it intended to create an attorney-client privilege. If you have legal questions, please consult your personal attorney. Any prescriptions described in this book are for informational and educational purposes only. They are not individual prescriptions for any medication for any person. The authors and the publisher cannot take medical or legal responsibility of having the contents of this book considered as a prescription for everyone. The authors and publisher have neither liability nor responsibility to any person or entity with respect to any loss, damage, or injury caused or alleged to be caused by the information contained in this book.

AN EXTRAORDINARY POWER TO HEAL
Wisconsin Institute of Nutrition, LLP
http://www.nutritioninstitute.com

copyright (c) 2003 by Bruce Semon, M.D., Ph.D. and Lori Kornblum

All rights reserved, including rights of duplication and republication in whole or in part, including storage in a retrieval system or transmission in any form by any means, electronic, mechanical, photocopy, recording or otherwise, without written permission of the authors, except for including quotations in a review or personal use or as specifically noted in certain appendices.

Cover design and graphic consultation: Tiffany Navins, Graphic Ingenuity, Germantown, Wisconsin
Cover art: Jeanine Semon, http://www.jeaninesdream.com
Book design: Lori Kornblum

Library of Congress Catalog in Publication Data:

Semon, Bruce, M.D., Ph.D., and Lori Kornblum
    An Extraordinary Power to Heal
    Includes index
    1. Candida Related Complex   2. Candida Diet
    3. Autistic Disorder   4. Psoriasis
    5. Attention Deficit Hyperactivity Disorder
    6. Multiple Sclerosis 7. Tourette's Syndrome 8. Ulcerative Colitis
    9. Allergies 10. Otitis Media

Library of Congress Control Number: 2003106367
ISBN 0-9670057-4-4
First Edition   September, 2003   10 9 8 7 6 5 4 3 2 1

To order single copies or to schedule author appearances, contact us at 1-877-332-7899, or http://www.nutritioninstitute.com. email: bsemon@nutritioninstitute.com.

About the Authors

Dr. Semon and Ms. Kornblum are the authors of *Feast Without Yeast:4 Stages to Better Health* (1999) and *Extraordinary Foods for the Everyday Kitchen* (2003), published together with this book. *Feast Without Yeast* has sold thousands of copies and is available throughout the world.

Dr. Bruce Semon is board certified psychiatrist and child psychiatrist, as well as a doctorate level nutritionist, practicing in Milwaukee, Wisconsin. He received his M.D. from University of Wisconsin-Madison and his Ph.D. in Nutrition from University of California-Davis. Dr. Semon was a Research Fellow in the Laboratory of Nutritional and Molecular Regulation of the National Cancer Institute at the National Institutes of Health. He received his adult and child psychiatry training at the Medical College of Wisconsin.

Dr. Semon has published several academic papers relating to nutrition, and is a contributing author to *Biological Treatments for Autism and PDD*, by Dr. William Shaw.

As a complement to Dr. Semon's regular psychiatry practice, Dr. Semon has treated many patients for yeast-related illnesses, including all of the medical conditions described in this book, with remarkable results. Dr. Semon speaks regularly to support groups and conferences about the connection between what people eat and their health. Dr. Semon accepts new patients.

Lori Kornblum is an attorney practicing in Milwaukee, Wisconsin. She has published scholarly and popular articles relating to law. Ms. Kornblum has taught cooking classes to implement special diets and speaks about how to change diet.

Dr. Semon and Ms. Kornblum are married and live in Milwaukee with their three children, one of whom has an autistic disorder.

Dr. Semon and Ms. Kornblum both are available for speaking engagements and consultations. For more information about becoming a patient of Dr. Semon, or for information about speaking engagements or consultations, please call toll free 1-877-332-7899, or write to us at support@nutritioninstitute.com.

Acknowledgments

We gratefully acknowledge the assistance of Catherine Lane, E'liane Khang, and Susan Brodie, Ph.D., for their assistance in editing this book. We also acknowledge the assistance of our graphic artist, Tiffany Navins, Graphic Ingenuity, for her creative input, and artist Jeanine Semon for her inspirational artwork.

We expecially acknowledge the patients whose life experiences are chronicled in this book. These patients courageously departed from standard medical treatment, which did not work for their conditions, and tried a different treatment that we describe in this book. They faced challenges from friends, family and even their own doctors, but persevered and were rewarded with health that they had not experienced, in many cases, for years. While we cannot acknowledge each patient individually, to protect their privacy, we thank them.

I, Dr. Semon, would like to acknowledge the teachers who taught me how to research and understand medicine by questioning and not simply accepting things as they are. These outstanding teachers are Dr. Richard Proctor, University of Wisconsin-Madison; the late Dr. Philip M.B. Leung, University of California-Davis; Dr. James Phang, National Cancer Institute at National Institutes of Health; and Dr. Richard Freedland, University of California-Davis.

Finally, we acknowledge the pioneering work of Dr. Orian Truss, Dr. William Crook, Dr. Sidney Baker, Dr. William Shaw, and Dr. Bernard Rimland.

This book is dedicated to our children,
Avi, Sarah and Mikah
and to the memory of
Philip M.B. Leung, Ph.D.

Contents

PART I
Introduction--How What We Eat Affects Our Health

Chapter 1

An Extraordinary Power to Heal: What this Book is About

My doctor told me that nothing else could be done for me and that I would have to live the rest of my life this way. Isn't there something else that can be done?

Many of my patients come to me with this question, desperate to find help that they could not find from their other doctors.

Fortunately, I have been able to tell them that the short answer to their question is "yes." There is a treatment that will help. This treatment is the subject of the rest of this book. You will see cases of my patients throughout this book, and find out that the treatment I prescribe has helped real people. Only their names have been changed to protect their privacy.

Many people have experienced the frustration of being told nothing could be done to help them feel better. They had a medical condition that was painful or uncomfortable, had been to numerous doctors, and had been told that no therapy was available. The doctor might have said that all the patient could do was "learn to live with it".

However, much more can be done for these many hurting and disappointed people. This book is about a different way of treating many medical conditions, some considered incurable. The treatment is nontoxic and not expensive. Almost everyone who has followed through on this treatment has improved significantly. The treatment consists of three parts. First, you take all foods out of your diet that cause you to be unhealthy. I tell you what those foods are in Chapter 2, and tell you how to eliminate them in Chapter 16. Second, you add good foods to your diet to improve your health. I tell you what those foods are in Chapter 16. Third, you need to kill the yeast Candida albicans already in your body. I tell you how to do this in Chapter 17.

Candida albicans is a single-celled micro-organism called a "yeast" which can attach itself to your inner intestinal lining as well as other places in your body. There it can grow and reproduce and cause tremendous harm to your health. Candida albicans is also found in the soil, in the air and in food. This book is about the harm that comes from having too much Candida albicans in your body and how you can eliminate it and feel healthy.

Who can benefit from the treatments described in this book? Based on my clinical experience with patients I personally have treated, anyone with any of these conditions will benefit from anti-yeast treatment.

- **acne**
- **allergies**
- **anorexia and eating disorders**
- **asthma**
- **attention deficit hyperactivity disorder**
- **autism**

- autoimmune disorders such as:
    - multiple sclerosis

        rheumatoid arthritis

        Crohn's disease

        ulcerative colitis, and others
- candidiasis
- chronic abdominal pain
- chronic diarrhea
- chronic ear infections
- chronic fatigue
- chronic sinus infections
- chronic vaginal yeast infections
- chemical sensitivity
- constipation
- decreased sex drive
- depression and other psychiatric disorders
- endometriosis
- food cravings, especially for sugar and chocolate
- gas (flatulence)
- loss of motor ability and/or speech
- migraines and other headaches
- obesity
- poor concentration
- recurrent miscarriage
- skin disorders, including:
    - psoriasis

        eczema

        generalized itching

- **speech loss**
- **tactile sensitivity**
- **Tourette's Syndrome**

What are my qualifications to write this book? I hold both a major research degree, a Ph.D. in nutrition from the University of California-Davis, and I hold an M.D. (medical doctor) degree from University of Wisconsin (Madison). I also spent two years studying diet and cancer prevention at the National Cancer Institute in Maryland. I am a board certified psychiatrist. I have treated all of the above conditions using the program described in this book.

In a Chapter 19, I will describe reasons why the traditional medical research community finds the study of Candida yeast so challenging. The combination that I possess of a knowledge of medicine and research as well as a thorough understanding of how our diets influence our health, has been essential to my ability to treat Candida yeast.

Let me give a case example of how the treatment described in this book helped Heidi, a young child who had extreme anger and developmental problems.

Heidi

Heidi, at 3 and a half years old, came to see me because she had uncontrolled anger, violent tantrums, and developmental problems. She was hitting other children at nursery school and hitting her parents. In the past she had bitten herself. She would throw herself on the floor and hit her head, to the point where she would bruise her head. She would refuse to go to bed and would throw tantrums at bedtime or when food was taken away. These tantrums had begun a year prior to the appointment. Heidi also threw tantrums when she was told "no" or was not getting what she wanted. She would throw tantrums if certain things were not in place or if her clothes were not arranged properly. She liked to blame others for her tantrums.

Even at age three and a half, Heidi was talking about death. Her family had experienced several deaths. In addition to her problem with tantrums, Heidi had significant sleeping problems. She could take two to three hours to fall asleep and was sometimes waking during the night.

Heidi could speak and was intelligent, but she nearly became aggressive with me.

A psychologist had previously labeled Heidi "strong willed." The psychologist's solution was to blame the parents by telling them to apply more consistent discipline.

I did not view Heidi's behavior as a parenting issue. In fact, Heidi's symptoms were also symptoms of potentially long term developmental disabilities, including autism (a disorder in which children are not able to interact appropriately, and may become locked into certain rituals). The possibility of autism was very serious. In my view, Heidi needed help and treatment.

I suggested to Heidi's parents that they try an approach of changing Heidi's foods to my 4 Stages, described in Chapter 16, an anti-yeast food plan, and giving the anti-yeast medicine, nystatin, the use of which I describe in Chapter 17. These therapies are nontoxic. Heidi's parents first started her on melatonin, a natural hormone for sleep, and then on the 4 Stages food plan.

After one week on the food plan, Heidi was sleeping much better. She was having fewer tantrums and the tantrums she had were not violent. Her parents began giving nystatin after the second appointment.

When I saw Heidi a month later, her aggressive and self-injurious behavior had stopped. She was sleeping well and she was interacting appropriately. Her parents told me that she was a totally changed child for the better. Heidi told me proudly with a big smile that she had helped decorate the Christmas tree.

This case shows how a child with severe tantrums and early signs of developmental problems was put back on track in five weeks. This case was not an example of a parenting issue.

Like Heidi's parents, you may have been told that your medical condition or that of your child was a behavioral problem, or was incurable, or that the only therapy available was medication or surgery, or that there was no solution and you would just have to "learn to live with it." I learned in medical school and postgraduate training that no one knows the cause and cure of all the disorders I describe in this book. In most cases, all the doctor can offer is symptomatic relief. When people suffer from autoimmune disorders, such as rheumatoid arthritis, in which the body's immune system attacks the body's own organs, the pain can be so awful that patients endure very toxic drugs with numerous side effects just to have some relief.

Modern medicine's inability to understand the causes of these illnesses can be very hard on patients and their families. Sometimes doctors cannot offer even much symptomatic relief.

However, another therapy is available.

What is this therapy? It is changing the food you eat and taking a nontoxic medicine to kill the intestinal yeast Candida albicans. In my clinical experience over more than a decade in practice, I have prescribed this treatment for all the conditions I describe in this book. I will briefly say how I came to learn about this treatment.

In 1990, my then four year old son was rapidly losing all of his ability to function. He was losing his speech and his ability to use his hands. The many doctors who were consulted could not help him. Although I am a medical doctor, initially I could not help my son.

When my son had lost almost all function, my wife figured out that he was having terrible chronic migraine headaches. To this day, I do not know how he stood up each day. The doctors did not believe us, mainly because they did not identify the problem, but we started to eliminate foods known to cause headaches. And he got better. He was more comfortable and his behavior improved. The improvement was visible within the first week.

The tremendous need to find something to help my son led me to the question of why eliminating such foods had caused such a significant change in him. I already knew that diet is important to our health. I had studied parts of the diet to earn my Ph.D. in nutrition. However, I had to delve more deeply into studying diet, and about the role of the yeast Candida albicans in causing illness. I developed a treatment technique which helped my son and which also has helped many other people over the last several years.

This treatment technique relies on significantly changing food choices and taking a nontoxic anti-yeast medicine, nystatin. The combination of these two therapies significantly relieves all of the conditions listed above, almost without fail.

How do you know if you may benefit from anti-yeast treatment?

Incredibly, the intestinal yeast Candida albicans may cause or make worse conditions ranging from psoriasis and fibromyalgia, to multiple sclerosis and chronic abdominal pain and the other conditions described in this book. One indication that yeast may be a problem for you is that your condition worsened after you took antibiotics. Antibiotics clear out intestinal bacteria and make room for the yeast. Any health condition which started or worsened after taking antibiotics is likely to be yeast related.

For women, taking the birth control pill or being pregnant also can make yeast worse. Yeast seem to do better when there is progesterone around. Progesterone is the hormone of pregnancy, and some form of progesterone is in birth control pills.

Think about what makes your problem worse or when the problem became worse. If your condition worsened after taking antibiotics or after a pregnancy, or after taking birth control pills, then yeast is likely to be a problem.

However, some people may suffer from problems related to yeast even if they have never taken antibiotics or been pregnant or taken birth control pills. The reason is that the food we eat contains chemicals and growth factors which help yeast to grow. All of us have some yeast. Whether a person with yeast suffers from a health problem depends, I suspect, on genetic susceptibility. Some people are more susceptible than others. We don't really know why.

Yeast can cause health problems by a variety of mechanisms which will be explained below and in the coming chapters.

For now, let me explain some basics about yeast.

What is this "Candida" yeast anyway?

Candida albicans is termed a yeast, a yeast-like fungus and an imperfect fungus,[1] which is found in the intestinal tract of human beings. Yeast cells are about the same size as our own cells. Unlike our cells, yeast cells have a capsule. This is an outer covering which I will describe in much greater detail in Chapter 8. Yeast, like our own cells, display "receptors." Receptors are branch-like structures extending from the surface of the cells. In the intestinal tract, yeast share space with much smaller microorganisms called bacteria. Bacteria can adhere, or attach themselves to, the inside wall of the intestinal tract. The good bacteria which adhere protect us from the adherence of disease causing bacteria, like Salmonella and Shigella, which cause diarrhea. Yeast unfortunately also can adhere to the inner intestinal wall.

The usual bacteria attached to the inner intestinal wall are benign and do not harm us. These bacteria do not make harmful chemicals or provoke immune responses and inflammation. When you take antibiotics for an infection, the antibiotics not only kill the bacteria causing the infection but also can kill and clear out these more benign bacteria in your gut. This loss of benign bacteria makes room for yeast and other disease causing bacteria.

Although most people regard a significant amount of Candida as normal, this is only because many people have significant amounts of yeast. What is "normal" is not necessarily "good." The intestinal yeast Candida albicans is capable of and frequently does cause major health problems by a number of mechanisms. I explain each of these mechanisms thoroughly in coming chapters. For now, please note that yeast works in a variety of ways to cause devastating problems.

Does this yeast do any good?

People have asked me if this yeast does anything positive. The short answer is no. To use an analogy, if this yeast were your neighbor, you would be living next to a leaky toxic chemical plant.

Fortunately, a defense against this noxious neighbor exists. What is it? Keep on reading. Let me first give you a historical perspective of yeast related medical findings.

A brief history of yeast-related medical observations

The concept that yeast in the body can cause health problems ranging from diarrhea and depression to multiple sclerosis and rheumatoid arthritis was first developed by Dr. Orian Truss. Dr. Truss, an allergist in Birmingham, Alabama, describes in his book *The Missing Diagnosis* how he found that patients with problems such as depression actually suffered from yeast.2 In 1961 he had a female patient suffering from allergies, migraine headaches, depression, and premenstrual syndrome. She was allergic to the yeast Candida albicans. As allergists do, he gave her a very dilute extract of the substance to which she was allergic, the yeast Candida albicans. Giving this dilute extract apparently reduced the Candida in her body. Dr. Truss found that not only did his patient's headaches resolve, but so did her depression. He theorized that the yeast was making some toxic product which was then absorbed, causing the brain and other parts of the body to suffer disease. Dr. Truss followed up his theories with practice. He treated many people who suffered from depression to sinus problems by treating their yeast allergies.

Dr. Truss' findings have led to the publication of many books on the successes of treating health problems by looking at the yeast as the cause of those problems. Dr. William Crook, a pediatrician in Tennessee, has spent years collecting the experiences of anti-yeast practitioners and publishing these experiences in several books, including *The Yeast Connection*, *The Yeast Connection Handbook*, and many other related books, with cases of people who got better when their yeast was treated.3 Dr. Crook's publications, talks and other efforts have helped many people become aware about the problems yeast can cause.

Why we wrote this book: To make anti-yeast therapy more understandable, more effective and more accessible

I have found, as a doctor and as the parent of a patient, that people need more than what currently has been available. People needed a more accessible and easy to follow food plan, so my wife and I wrote *Feast Without Yeast: 4 Stages to Better Health*.4 *Feast Without Yeast* presents a very easy to follow diet and more than 225 original, yeast free recipes. Almost 200 of those recipes also are free of wheat, dairy and eggs. *Feast Without Yeast* did not answer all of the questions people had, however. I found that even more than recipes, people needed more scientific and clinical information about why anti-yeast treatment works, so they could go to their own doctors for treatment.

We embarked on writing *An Extraordinary Power to Heal* to fill the gap between what patients know works and convincing their doctors to help them implement this plan. *An Extraordinary Power to Heal* explains in language that both you and your doctor can understand why anti-yeast treatment is so effective. The footnotes in this book give additional references so you and your doctor can look up the studies on which we rely.

In *An Extraordinary Power to Heal*, I record and explain my experiences with treating patients who have the intestinal yeast Candida albicans. More than that, this book gives you the scientific background to understand why anti-yeast treatment works. This is a book you can read, understand and take to your doctor. As you will see, even problems which the medical community considers incurable can be treated by removing Candida albicans from your intestinal tract. The results of such therapy can be quite amazing. My experience is that, for the problems described in this book, no traditional medical therapy comes close to being able to do for people what anti-yeast treatment can do.

I want you to be able to understand what anti-yeast therapy can do for you.

What is the experience of patients who follow different anti-yeast treatments?

Many patients have read books on yeast and its treatment and have not had the same successful experience. The books may have helped them understand that yeast was a problem, but the treatments described in other books are harder to follow. When the patients approached their doctors for the suggested medications, their doctors often would not prescribe them. This experience is a recurring theme in conversations with people who call, e-mail or who come to see me. It is very frustrating for patients.

I wrote this book so both parts would be easier, the food plan and the medications. The food plan, the 4 Stages diet, is described in this book in great detail in Chapter 16. You can do this yourself. The medical plan suggests a prescription medication called nystatin. Chapter 17 is about nystatin. Appendix C gives you complete information about how to start nystatin. You do not need to wait for a prescription for nystatin to start the food plan.

This book provides information to share with your doctor

I have found that for the best anti-yeast treatment, patients should take the non-absorbed medication nystatin. Nystatin kills yeast in the intestinal tract. Unfortunately, many doctors are not interested in treating intestinal yeast.

The medical establishment looks for answers in certain places such as medications, in medical devices, in surgery and now in human genes. Food and intestinal yeast are not high up on the priority list. Medicine also looks at health problems by the area of the body or system affected. Most doctors are not trained or accustomed to look at how diet affects health nor are most doctors trained to look at classes of medical conditions that cross traditional boundaries.

To convince your own doctor to prescribe nystatin at all and then for a long enough time, can be difficult. Of course, you could seek another doctor, like myself who specializes in treating medical conditions that respond to anti-yeast treatment. However, I have found that most patients would prefer to convince their own doctors to provide treatment. When people ask me, "Can my doctor do this for me?," the answer is yes, if the doctor would like to do so. The doctor need only be willing to write a prescription for nystatin and supervise your treatment with the 4 Stages diet.

I wrote this book so your doctor will understand from a medical perspective why this treatment works and why the risks of adverse effects are so minimal. As a doctor myself, I understand the kinds of questions your doctor will ask. I answer those questions in this book. I wrote this book so your own doctor can understand how to help and treat you.

I also wrote this book for you. I find that patients who understand the nature of their problems and treatment do better. And I want *you* to feel better! With this book, you can take charge of your own health and talk with your doctor. Of course, if your doctor remains unconvinced and you want medical treatment, you can see me or another doctor who understands this treatment.

This book presents case studies of patients that I have treated successfully. I hope these case studies help you and your doctor by showing how many different patients and with many different medical conditions benefitted from anti-yeast therapy. The explanations give insight into why such improvement occurred. The details of the treatment are given.

Why is the treatment described in this book more effective than treatments described in other books?

What is the difference between this treatment and that recommended by others? The difference is that the types of food changes that I recommend are different from the anti-yeast diet that Dr. Truss originally recommended, and which most others follow. The typical anti-yeast diet is high in protein and low in carbohydrate. I have found that this diet is not the most effective diet for fighting yeast.

The diet that I developed, called the 4 Stages diet, is very different from the standard anti-yeast diet. I will give you an overview here but will give you detailed directions about how to implement the diet in Chapter 16.

To understand why the 4 Stages diet works for the problem of Candida albicans, you need to understand the concept that yeast make chemicals which kill bacteria. These chemicals make room for the yeast by killing the bacteria that share space with the yeast. Many of our most common foods also contain such chemicals. I discuss these problems thoroughly in Chapter 2. For now, you just need to be aware that to clear the yeast, you need to limit your intake of these chemicals. Otherwise the medicines used to kill yeast don't help much. If you do not change your food choices, the dietary chemicals will kill the bacteria and enable the yeast to grow back. Yeast treatment depends on the right diet. Such changes of food choice can be very powerful. Many patients have become better from these food choice changes alone. The therapy works better with nystatin, but the therapy starts with the change of food choices. More information on food will be presented in the next chapter.

The original high protein, low carbohydrate diet relies usually on eating more meat. Candida can still grow on meat. Other problems with this dietary approach will be explained in the next chapter.

Conclusion

Now that you have had a very brief introduction to how yeast can affect your health, let us move forward to see exactly what this problem is all about and how it is treated.

Notes

[1]A true fungus produces branch like structures called hyphae and can reproduce either sexually or asexually. An imperfect fungus cannot reproduce sexually. In the case of Candida, the branch like structures are not exactly the same as those of a true fungus, so they are called pseudohyphae and Candida can only reproduce asexually. In this book, Candida will be referred to as a yeast, but it could also be called an imperfect fungus. The common bread yeast is called Saccharomyces, and this yeast does not usually grow inside of humans. Candida and Saccharomyces both make chemicals such as alcohols.

[2]Truss, C. Orian, *The Missing Diagnosis*, 1982, C. Orian Truss, P.O. Box 26508, Birmingham, Alabama, 35226.

[3]Crook, William G. *The Yeast Connection Handbook*, 1999: Professional Books, Inc., Box 3246, Jackson, TN 38303.

Crook, William G. *The Yeast Connection, A Medical Breakthrough*, originally published by Professional Books, Inc., Box 3246, Jackson, TN 38303, 1983. This book was later published by Random House, Inc., New York.

[4]Semon, B. A. and L. S. Kornblum. *Feast Without Yeast: 4 Stages to Better Health.* 1999: Wisconsin Institute of Nutrition. For further information, see http://www.nutritioninstitute.com, or call 1-877-332-7899.

Chapter 2

Food Choices, Yeast, and Your Health

I could not have written this book if I were not interested in what is in food and how food affects health. Food profoundly affects health and well-being, both positively and negatively. The most important positive way food affects health is that eating the right foods and avoiding the wrong ones makes you healthier. One of the most critical negative ways that food affects health is by helping yeast grow in the intestinal tract. This chapter explains how food helps yeast grow in your intestinal tract, and why this is bad for your health. I will tell you what you can do to solve this problem.

For example, one of the worst foods for you is malt. Malt is a sugar substitute which is baked into nearly every commercial product, ranging from breakfast cereals to bagels to breads. Malt contains twenty chemicals that put your brain to sleep. Anyone who eats malt in the morning will have problems concentrating, because the malt will be putting his or her brain to sleep. Should we wonder why so many children cannot concentrate in the morning?

Malt also contains the major headache causing chemical tyramine. How many people get headaches at school or at work?

Malt is produced by sprouting a high grade barley, then heating the sprouting barley. Malt is specially raised to feed yeast for making beer. Malt also feeds the yeast in your intestinal tract and helps the yeast grow. Yeast make all kinds of toxic chemicals. These toxic chemicals affect your health. I explain how below. So one easy, but very major step you can take to improve your health is to eliminate malt.

The information which I have just given you about malt, for example, is not available unless you know very specifically where to look. Yet what is in your food affects your family's health every day.

I wrote this book to make such information more available.

Food choices for fighting intestinal yeast

Intestinal yeast diminishes health and well-being. Making the right food choices is exceedingly important in fighting the yeast in your intestine. In this chapter, I will explain which foods hurt your health by helping the yeast Candida flourish in your intestinal tract. Many people immediately assume that the worst food to eat is yeast itself, as in bread. Bread yeast is not the best thing to eat but it is not the worst either. Saccharomyces is the common bread yeast and is the most common industrial yeast. The baking process kills this yeast. The yeast in your intestines is Candida albicans, a different yeast. You will see in this chapter that I am discussing different types of food, many of which will surprise you.

To prevent something from growing, the first thing you want to do is eliminate its nutrient sources. Without nutrients, an organism cannot grow. Yeast grows in your intestinal tract, where your food is digested. Yeast lives off things that you eat. So to prevent yeast from growing in your intestinal tract, we should take away from the yeast everything in food which helps them grow.

Although yeast can use just about any food for growth, some foods support yeast growth much better than others. You want to eliminate the foods which make it easy for yeast to grow. I wrote this book to give you knowledge about how to eliminate the right foods. The food which best supports yeast growth is malt. Why? Malt is raised specifically to feed the yeast as it makes beer. Malt does the same in your intestine. Nothing in your food helps yeast grow faster. So taking out malt substantially reduces yeast growth rate.

Some foods must be eliminated from the diet because they contain chemicals which kill bacteria and make room for the yeast. As will be discussed below, yeast and bacteria compete for space in your intestinal tract. Chemicals which kill bacteria give the yeast more space and an advantage for growth. To keep the yeast growth down, you must eliminate foods which contain chemicals which kill bacteria and give yeast a growth advantage.

Another group of foods to eliminate are those which mimic yeast's bad effects. These foods contain chemicals similar to the bad chemicals that yeast make which hurt your body.

Many people eat these foods every day, and could benefit from eliminating even some of these foods part of the time. For example, ketchup, mustard, and even most commercial white flour is bad for you. You will see what the rest of the foods are as you read this chapter!

Antibiotics, your food and yeast

Antibiotics kill much intestinal bacteria and make large amounts of room for the yeast to grow. Your food contains chemicals which kill bacteria and also make room for the yeast to grow. You can get yeast problems from antibiotics, from eating the wrong foods or both. Antibiotics are powerful at causing yeast problems. However, antibiotics are given only for illnesses. They should not be part of your everyday intake. Food is eaten every day. Foods containing anti-bacterial chemicals are important because every time such foods are eaten, they clear out bacteria and help the yeast. For example, vinegar is one of these foods. You will find more as you read this book. Collectively, the effect of food is large because food is eaten every day. Eating the wrong foods hurts and eating the right foods helps.

Why not just use an anti-yeast drug?

Before discussing which foods to eliminate, we need to explore why making proper food choices is important, and why you cannot just kill the yeast by using an anti-yeast drug. Many people ask, "why can't I just take an anti-yeast medicine to kill the yeast? The diet seems awfully inconvenient." People are used to going to their doctor and coming away with a simple prescription that will solve all their problems. Unfortunately, anti-yeast treatment does not work this way.

To understand why, we need to examine the intestinal tract. The intestinal tract is a long tube designed to help digest food and eliminate waste. Many organisms live inside your intestine. The ones that concern us are yeast and bacteria. Yeast share space with bacteria in the intestinal tract. Most of the bacteria are beneficial and help us digest food. We need to leave the bacteria alone so they can grow. This balance is disrupted when we take antibiotics to kill infectious bacteria. Taking these medicines leaves space in the intestine for the yeast to grow. Foods in the diet significantly influence how the yeast and bacteria will regrow in the intestine after medications have been given. The food choices you make, what nutritionists call "diet", is extremely important in enabling the "good" bacteria, and not yeast, to come back. Remember, the food you eat affects the bacteria, yeast and how they grow, every day.

The average American diet contains chemicals and growth factors which will favor yeast growth and keep the bacteria down. The average American diet can undo the beneficial effects of even the best anti-yeast drugs, and will help yeast grow back even after using of such medicines. From my clinical experience, anti-yeast drugs cannot help in the long run unless food choices are changed.

The best approach to reducing intestinal yeast is to exclude what helps the yeast to grow

The best approach to reduce intestinal yeast is to exclude foods containing anti-bacterial chemicals and foods containing growth factors for yeast. I will discuss these types of foods and the reasons for eliminating them in this chapter. I will also discuss foods that contain toxic yeast chemicals. Yeast chemicals are poisonous or toxic, directly to bacteria. Such chemicals are also poisonous to human beings. These chemicals also cause headaches, and may cause other problems. Another way in which toxic yeast chemicals are harmful is to sedate and slow down the brain and nerves.

Yeast and Bacteria Compete for Space

Before getting to the list of foods which you should avoid because they contain yeast growth factors, or because they contain chemicals which kill bacteria, we need to look at some basic principles of how fungi, yeast and bacteria compete against each other.

In the natural world, microorganisms such as fungi, yeast and bacteria all inhabit the same places and compete for space and nutrients. Yeast and bacteria can be found in the same places in the soil, in and on plants and animals, and other places. What keeps these microorganisms from overrunning each other in nature? Because yeast, fungi and bacteria all want the same space to grow, and the same nutrients, they make chemicals to kill each other. Yeast, mold, and fungi naturally produce anti-bacterial chemicals as they grow, such as penicillin and alcohols. Yeast and mold make these chemicals because they want to keep the bacteria from utilizing the same nutrients as themselves. For example, as a certain mold grows, it makes penicillin to kill bacteria. The mold or fungus is growing on a nutrient source, such as bread. As the mold grows and makes penicillin, the penicillin kills the bacteria. Then the bacteria cannot utilize the nutrients of the bread to grow, leaving more nutrients for the mold and fungus.

Some bacteria are not defenseless. As the yeast and fungi make chemicals to kill bacteria, some bacteria make chemicals to kill yeast and fungi. Two of these natural antifungal chemicals are amphotericin and nystatin. These names may sound familiar because both are antifungal medications, which doctors prescribe to kill yeast and fungi.

One of the places you can find the yeast Candida albicans and bacteria side by side is in your intestines, or gut. The main yeast in your gut is a species called Candida albicans. Candida albicans and bacteria both try to attach themselves to your inner gut lining. Bacteria grow much faster than yeast, and theoretically could overgrow the yeast. But the bacteria cannot do this if there is significant yeast present. Why? The yeast make anti-bacterial chemicals such as alcohols.[1] The alcohols will kill bacteria, making more room for the yeast. The usual bacteria in the gut are beneficial to your body, and are not known to produce any toxic chemicals to kill the yeast.

As food comes into your gut, it digests the food, releasing nutrients. Then the yeast and bacteria use these nutrients from the food. If you eat food containing chemicals which kill bacteria, you will give the yeast more room to grow. If you eat foods such as malt which contain very specific growth factors for yeast, the yeast will grow much better. If you avoid such foods, you will help the beneficial bacteria grow. The basic principle behind the 4 Stages diet, which I explain in detail in this chapter and Chapter 16, is to remove the anti-bacterial chemicals and yeast growth factors coming in from the diet, to inhibit the yeast and give the beneficial bacteria a better chance to grow.

Yeast make toxic anti-bacterial chemicals

Let us look at how yeast makes anti-bacterial chemicals. Fermentation is the process by which any type of yeast is added to a food and allowed to grow in the food. The result of fermentation is, for us, often seemingly good tasting food. Two examples are bread making and beer brewing. Other fermented foods are chocolate and soy sauce. When beer brewers add yeast to malt, the yeast digests the malt and forms ethanol and other alcohols and chemicals. The result is beer. When wine makers add yeast to grapes, they get wine. When manufacturers add yeast to cucumbers, they get pickles. During pickling and fermentation, the yeast makes chemicals to kill bacteria. These chemicals preserve the food by preventing bacteria from coming in. Otherwise, bacteria would use the food's nutrients and spoil the food.

People use fermentation to preserve foods. Many people like some of these yeast chemicals, such as ethyl alcohol, the primary alcohol found in alcoholic beverages. However, as I will discuss below, some of the chemicals synthesized during fermentation are poisonous.

Eating toxic chemicals harms your health

The yeast chemicals made during fermentation which kill bacteria do not disappear when they enter your mouth. When people eat or drink fermented foods as part of their diet, they also eat or drink chemicals which kill bacteria. These chemicals continue to act in your body and gut, clearing out beneficial bacteria from your gut and making room for yeast. When such anti-bacterial chemicals are present in your food, they help intestinal yeast win the battle against the intestinal bacteria. Then yeast grows better in your intestinal tract. The beneficial bacteria will not be able to grow if you eat chemicals to kill them. Getting rid of the yeast in your intestinal tract without first excluding these anti-bacterial chemicals is very difficult. So what foods help yeast the most?

Health Enemy #1: Vinegar

Anything which has been subjected to yeast or fungus during processing potentially has anti-bacterial chemicals. Wine, beer and other alcoholic beverages have anti-bacterial chemicals. However, many other such foods exist, some almost hidden. The most important of these foods is vinegar, because it has so many harmful anti-bacterial chemicals. One is ethyl acetate, which is present in large quantity.[2] Ethyl acetate kills bacteria,[3] and is used in industrial food processing, such as bread making, to keep the bacteria down without touching the yeast. In fact, ethyl acetate kills bacteria at the same time that it leaves yeast alone.[4] So eating vinegar increases your intake of chemicals which kill bacteria, helping yeast to flourish in your intestine. In addition, vinegar also contains a number of toxic alcohols including methanol.[5] These alcohols are toxic to your brain and also kill bacteria.

You can find vinegar as a major ingredient in many common foods, including ketchup, mustard, salad dressing, bread, and mayonnaise. You must read food labels to find all the vinegar in your diet. Eating any of these vinegar containing foods will help yeast grow in your intestine. Vinegar is your #1 health enemy. This is the first food to eliminate on a yeast-free diet.

Health Enemy #2: Malt

Malt is as bad or worse than vinegar, and like vinegar, is found in many foods. Malt is a seemingly harmless sugar substitute which is wreaking havoc, especially on children. Malt contains growth factors for yeast. You must eliminate malt to clear intestinal yeast.

What is malt? Malt is a processed food that starts from a

special barley. The barley is sprouted, then heated. The resulting product is malt. It is also called malt barley, barley malt, malt extract or malt syrup. Barley malt is the raw material for beer. Malt also is sweet and flavorful. You can also find it in health food stores sold as a grain-based sugar substitute. Don't be fooled! Table sugar is better for you than malt. Why is malt so terrible? After all, many of us grew up drinking malted milks. Malt contains more than twenty toxic chemicals. Malt contains major growth factors for yeast.

Malt's composition is complex. We get the best idea of why malt causes so many problems by looking at how industry uses it. The main use of malt is to grow yeast in large quantities. Malt is used to make beer. Yeast is added to the malt, and the yeast makes beer. Beer making is competitive and consumes large quantities of yeast. Nothing helps the beer brewers generate more yeast than malt. Bakeries also use malt. Most commercial white flour contains malt. Here again, bakers need to help the yeast grow.

To see how well malt helps yeast grow, we can look at laboratory studies. Scientists actually use malt to test whether yeast is present. A sample containing an unknown amount of yeast is added to a malt containing culture to see if yeast grows. If there is any yeast in the unknown sample, the malt will help it grow so that the scientists can found out if yeast is present. In other words, the best way to grow yeast in the laboratory is to put the yeast on some malt.[6] This is because malt contains growth factors for yeast.[7] These laboratory observations are consistent with the fact that malt is used to feed the yeast as it makes beer and bread. You must eliminate malt to control intestinal yeast growth.

Malt may also suppress the body's immune system

Malt may do more damage than just helping yeast to grow. Malt may have significant effects on the body's immune system and its ability to fight Candida. Malt contains chains of two, three, or more sugar molecules.[8] These chemical structures are different from table sugar (sucrose), and from honey (fructose). Scientists have shown that such long chains of sugar molecules interfere with the interactions between immune cells.[9] If immune cells cannot communicate, they cannot fight infections, and they cannot fight yeast. Inhibiting such interactions results in decreased immune response in general. Inhibiting these interactions also results in decreased immune response to yeast. I discuss these interactions at greater length in the Chapter 8 about how Candida affects the immune system.

In other words, malt helps yeast grow and malt hurts your ability to fight yeast. This is why eliminating malt is so helpful to people suffering from yeast problems.

Malt causes other problems

Besides feeding yeast and suppressing the immune system, malt causes problems independent of yeast because malt contains more than twenty toxic chemicals. One of them, tyramine, found in large quantities in malt, causes headaches. Others, N-methyl tyramine and hordenine, are poisonous.[10] Eliminating malt helps decrease chronic headache problems. Twenty of the chemicals slow down the brain. No wonder kids cannot concentrate when they eat malt filled cereals for breakfast.

Now we have seen that health enemies #1 and #2 are vinegar and malt. In my clinical experience, most people who just eliminate vinegar and malt feel much better within a few days. In one case, a borderline hyperactive (ADHD) child eliminated vinegar and malt. His behavior improved so greatly that his parents will never go back to ketchup and other such foods. He went from being disruptive and unable to concentrate to being a normally active child who likes to sit and read.

In my clinical experience, eliminating vinegar and malt is the best way to start a yeast free diet. Many people, including the child I just mentioned, improve so much that little other therapy is needed.

For other people, however, vinegar and malt exclusion is a beginning. So we need to explore other health enemies.

Health Enemy #3: foods processed using fungus: soy sauce, chocolate, coffee and vanilla

Food processing is another source of yeast chemicals because food processors use fungus and yeast. The problems with food processing are much less well documented by laboratory studies, but I know from clinical experience that these foods cause major problems. Eliminating such foods helps people feel much better.

During processing, yeast and fungus generate many chemicals. Let us look at soy sauce. Soy sauce is a prime example of a processed food that hurts people. The process of making soy sauce is abhorrent to food scientists. Soybeans are mixed with fungus and yeast. The fungus and yeast grow, using the soybeans as food. There is no sterilization. Food processors use this same method to make miso and tempeh, two other soy-based foods.[11] The reason food scientists find this process abhorrent is that when fungus contaminates food processing, the food usually tastes terrible and is ruined. Fungus make very bad tasting, toxic chemicals.

The reason for discussing soy sauce here is that anything which fungus and yeast make will contain anti-bacterial chemicals. The fungus does not want any of the usual bacteria around. It is highly likely that soy sauce is anti-bacterial, and that it will kill bacteria and make room for yeast. Soy sauce therefore should be excluded from an anti-yeast diet.

Chocolate is also a product of yeast fermentation. Please do not stop reading here! I know there are many chocolate lovers out there who consider chocolate to be one of the four major food groups. Chocolate lovers should consider this.

Chocolate contains some of the same chemicals as malt. These chemicals are addictive. I will discuss the scientific basis of this addiction in Chapter 7. For now, let me say that the more chocolate you eat, the more you crave, in an endless cycle. Following the anti-yeast diet described in this book causes food cravings to disappear including chocolate cravings.

If you cannot imagine eliminating chocolate, start with eliminating other foods and taper off chocolate. You can test for yourself if chocolate is a problem by staying off of chocolate for a few weeks, then eating some. See how you feel, and see if you still think you need chocolate. There is life after chocolate.

We now return to food science. Chocolate is made by mixing cocoa beans with a variety of yeasts[12] and then placing them in open crates for four to nine days.[13] The cocoa beans are covered with banana leaves and allowed to ferment. The fermentation of coca beans is not exactly sterile. The fruit flies are all around. The yeast generate alcohol from the sugar in the beans. Chocolate may taste good but it is the result of fermentation, and has the same problems as soy sauce, beer and pickles. Fermentation results in yeast generation of anti-bacterial chemicals.

Coffee, like chocolate, is made by processing the raw coffee beans with yeast. Coffee beans are mixed with yeast to allow the yeast to break the beans down.[14]

Processing vanilla beans is slightly different, but it still uses aging and fermentation. First, vanilla beans are put in hot water. Then they sit out in the sun to "ripen."[15] This process takes two to eight weeks. Food scientists consider the primary process of ripening to be internal, inside the vanilla bean. However, this process is most likely a fermentation process analogous to that of cocoa beans, with yeast working on the vanilla beans while they are sitting in the sun. Vanilla extract is vanilla mixed with alcohol.

Worcestershire sauce is also aged, meaning it is fermented.

Processing foods by adding yeast and fungus creates the same types of anti-bacterial chemicals as found in vinegar.

Such chemicals will be found in foods such as soy sauce, tempeh, miso, chocolate, vanilla beans and coffee. These chemicals clear out bacteria and make room for the yeast Candida albicans.

You will feel better if you avoid these foods.

Health Enemy #4: foods contaminated by fungus and mold

The foods I just discussed cause problems because they are processed foods using fermentation, aging and/or yeast and fungus to create the finished product. Another category of foods that causes problems are the foods that contain significant amounts of fungal contamination. In nature, you will recall, fungus also make anti-bacterial chemicals.[16] Eating these anti-bacterial chemicals kills the beneficial gut bacteria, which makes room for the yeast. The most common fungus contaminated foods are cottonseed, peanuts, corn and apples. Cottonseed is contaminated with the fungus Aspergillus.[17] People don't eat plain cottonseed! But cottonseed is in many of our foods. You can find cottonseed oil in cooking oil and fat, especially in deep frying fats, used for commercial fast food. Cottonseed oil is used in many processed baked and cooked foods. Cottonseed oil is used for potato chips. If you start reading labels, you will find cottonseed almost as easily as you will find malt.

One place where you will find cottonseed that will surprise you is in chicken. Cottonseed is a staple of commercial chicken feed. Cottonseed is cheaper to feed to chickens than other feeds, so it is used frequently. One way animals and birds detoxify chemicals is by putting them into their fat. There the chemicals cannot do much harm to the animal or bird. The chemicals stay there when people eat the animal or bird. Then the people take in these fungal chemicals. The fungus chemicals will still be present in the chicken.

Farmers do not feed cottonseed to chickens which lay eggs. Eggs have much less if any fungal contamination.

Two other mold-contaminated foods are peanuts, which are frequently contaminated with the fungus Aspergillus,[18] and corn, which can be contaminated with fungus.

Two fungi, Penicillium and Aspergillus, also make an antibiotic compound, Patulin, which is often found contaminating apple juice.[19]   The fungus grows on the outside of the apple. To eat apples, peel first! I suggest avoiding apple juice. In the juicing process, the processors may not be so careful about which apples they use. They may use apples with more fungal growth and they may not remove the fungus completely in the peeling process. A child we knew suffered from terrible eczema on her arm. Nothing her mother tried solved the problem. I suggested eliminating apple juice. Within a few weeks, the eczema was gone and it never returned.

Antibiotics kill the beneficial gut bacteria, making room for yeast. Just think about all of the children in preschools around the country drinking gallons of apple juice to wash down malt filled crackers and cookies. These dietary snacks will cause yeast growth. We will see what yeast growth does to children's behavior in a later chapter.

The four main foods and food categories to avoid, then, are vinegar, malt, fermented foods, and mold and fungus contaminated foods. These foods contain chemicals which kill bacteria and make room for yeast. Some contain growth factors for yeast. Malt also contains chemicals that independently can cause health problems, like headaches and may affect your immune system.

I imagine that this information may seem overwhelming and a little hard to believe. How many of us grew up on vanilla ice cream and malted milks? Yet these foods and others are tremendously harmful for people suffering with yeast related problems, from multiple sclerosis to chronic fatigue. Eliminating these foods can free people from years of suffering. All of the foods I discussed above keep your and everyone else's intestinal yeast growing. All of these foods were discussed because they are the most concentrated sources in your diet of toxic and anti-bacterial yeast chemicals.

The choice between your health and fast food containing malt and vinegar may not seem apparent at the checkout counter of a fast food restaurant. Nonetheless, people make such choices every day, whether they are aware or not.

I just spent several pages discussing why you should eliminate certain foods because they help yeast grow. I have told you earlier that yeast is bad for your health. But how and why? What does yeast do to you? Much of the rest of this book is an answer to this question. Here we will start with toxic yeast chemicals. These chemicals put your brain and nerves to sleep. In other parts of this book, I will discuss how yeast also trick your immune system and cause major health problems.

When we look at chemicals that slow down the brain and nerves, the problem is more than yeast. Yeast make such chemicals but the food we eat also contains such chemicals, some of which are similar to the yeast chemicals. Many of these chemicals also kill bacteria. Beyond this, these chemicals slow down your brain and nerves.

Toxic yeast chemicals put your brain and nerves to sleep

Have you ever wondered why so many children and adults cannot concentrate well at school and at work? My experience as a doctor tells me that we should look for answers in the toxic sedative chemicals found in food and intestinal yeast. Let's go back to malt, my favorite example. As I said earlier, malt contains twenty chemicals which are toxic to the brain. These chemicals put the brain to sleep. How many people can stand up to those twenty chemicals? How many children are taking stimulant drugs to concentrate at school because the malt in their breakfast cereal put their brains to sleep? And malt is just one food that contains sedative chemicals.

Most of us would like to be awake and able to concentrate when we work. We would like our children to be awake and to be able to concentrate when they are in school. Most of us do not want to give our children drugs in the morning. However, when we eat the typical American diet, we are doomed to feeling tired as we try to concentrate. Our food contains many sedative chemicals, which combined with the intestinal yeast sedative chemicals, slow down the brain. You cannot achieve your goal of alert concentration when you're fighting against powerful chemicals in your body. These chemicals interfere with the ability to concentrate, both for adults and for children. I will list some of these chemicals and their sources so that you can avoid them. I will tell you about the chemicals coming from the yeast inside your intestines, as well as the chemicals from the food you eat.

Toxic sedative chemicals from yeast

A big source of toxic sedative chemicals is the yeast Candida albicans. Everyone has a small amount of Candida in their gut. If you take antibiotics, the amount of Candida increases markedly.[20]

Candida albicans produces a number of toxic chemicals which can then be absorbed into the body.[21] We know much about these toxic yeast chemicals because scientists study yeast and the chemicals yeast make during the process of making alcoholic beverages. These chemicals give alcoholic beverages a bad taste. The alcoholic beverage industry calls them "off flavoring agents". This industry studies yeast to determine how to get the most ethyl alcohol from the yeast, with the fewest bad tasting chemicals. However, even with significant amounts of ethyl alcohol present, the presence of other alcohols and these other chemicals can cause alcoholic beverages to taste bad. One of the reasons things taste bad is because they are poisonous.

Candida albicans (and other yeast) produce many toxic alcohols including ethyl alcohol or ethanol, found in alcoholic beverages, but Candida also produces much more poisonous alcohols.[22] The body does not clear these more toxic alcohols

quickly. Studies show the body clears them slowly.[23]
Scientific studies have shown that these toxic alcohols cause
major slowing and sedation of the brain, causing anesthesia,
narcosis, coma and ultimately death.[24]   Yeast also make
acetone,[25] which causes coma.[26] Yeast also make ethyl
acetate,[27] a chemical found in glue thinner. As do the other
chemicals, this chemical slows the brain down.[28]

In addition to alcohols, yeast make chemicals which are
poisonous to the body's production of energy.[29] With less
energy, the brain and body will not function as well.

This load of toxic chemicals will bring people down. No
one should be surprised that people who clear out their
intestinal yeast, by following the proper diet and taking
nystatin to kill yeast, feel better. Many patients who follow the
4 Stages diet have expressed to me that they feel more
energetic and can concentrate better.

Most people do not know that the yeast Candida make
these chemicals. The relative lack of knowledge has led to a
false sense of security about the impact of increased Candida
albicans in the intestinal tract.   Knowledge about these
chemicals would lead to greater concern on the part of doctors
about the impact of prescribed antibiotics.

Toxic chemicals from the bad bacteria Clostridia in your intestine

Antibiotics lead to more than yeast problems. Many people
with yeast problems also have an increase in a bad bacteria in
the gut called Clostridium difficile as well as other Clostridium
species.[30] Clostridium difficile synthesizes several very nasty
chemicals which are brain poisons that will slow down the
brain. These chemicals are listed in the footnotes.[31] Some of
the same dietary chemicals which favor yeast also favor
Clostridia. Otherwise, people who suffer from yeast and
Clostridia get into a vicious cycle. They take antibiotics to kill
the Clostridia which helps the yeast grow.

Sedative chemicals found in your diet

Our food also contains sedative chemicals. They slow the brain. Vinegar contains the sedative chemical ethyl acetate.[32] This is a sedative chemical which is found in glue thinner.[33] As I said earlier, vinegar is found in condiments such as ketchup and mustard, among other foods. Imagine a child eating a lunch of a hot dog on a bun (made with malt) with ketchup and mustard, with a pickle on the side. French fries cooked in cottonseed oil top off the meal. I cannot imagine how a child could learn in the afternoon. Their brain will only be partially present.

We have seen now that yeast makes toxic chemicals. These chemicals slow down the brain and cause other problems, including headaches. So we need to control yeast by eliminating foods that help yeast grow.

We also need to control the toxic load in our bodies by eliminating foods that contain toxic chemicals similar to yeast chemicals.

Let's look again at malt. Malt contains twenty toxic chemicals, called alkyl pyrazines, which sedate and slow down the brain.[34] The strongest of these chemicals is one-thirteenth as strong as phenobarbital.[35] That's pretty strong for a chemical found in food. Phenobarbital is prescribed in tiny amounts (60 to 90 milligrams) to induce sleep and prevent seizures. There are twenty such chemicals in malt. Malt is found in almost all commercial flour, and in most breakfast cereals. A child eating a standard bread and breakfast cereal in the morning is going to get a nice dose of malt and of alkyl pyrazines, making it hard to concentrate.

You can also find these same types of sedating chemicals, alkyl pyrazines in soy sauce,[36] chocolate,[37] potato chips,[38] and in roasted peanuts.[39]

In addition to the pyrazines, soy sauce contains a number of toxic chemicals.[40] These chemicals cause problems such as headaches and dizziness, and liver and kidney damage. I am not sure why soy sauce, which contains such toxic chemicals, is so desirable to cook with and eat, but I give one possible reason. Such chemicals will cause release of endorphins and this release is thought to be pleasurable. This topic is covered in the chapter on food addiction.

You can find another toxic chemical in cheese. The starter bacteria of cheese synthesizes the nerve poison hydrogen sulfide.[41]

Eating any of these foods will hurt you. Eating all of them will devastate your ability to concentrate.

Other toxic chemicals in your diet

Let me give some other examples of toxic chemicals in the diet. The fungi Penicillium roquefort, used to make blue cheese, makes Roquefortine A, B and C.[42] These three chemicals are all toxic. The most toxic one can cause death and other neurological problems. In other words, a fungus used to make a food product, Blue cheese, naturally makes other chemicals, which, in high enough quantities, are lethal.

I doubt most people would eat Blue cheese if they knew they were eating nerve poisons. But very few people know what is in our food. The popularity of Blue cheese shows how little knowledge is disseminated about what is in our food.

What are the effects of all these chemicals?

What is the effect of taking in such chemicals day after day, week after week? One effect is to slow down the brain, resulting in difficulty concentrating. The next main effect is to enhance the growth of Candida albicans, which makes its own toxic chemicals, which then slow down the brain as well as causing other problems that I describe in this book. These can be major health problems ranging from headaches to depression to Tourette's Syndrome, and other problems, depending on the individual. People who eat foods which yeast and fungi make will have more yeast in their intestinal tracts. We don't even know the full extent of the problems yeast cause.

You can see that I have discussed a great number of common foods. Most people don't just eat one of these foods. Remember the example I gave earlier, of the child eating a hot dog lunch. Imagine the child starts the day eating a breakfast of cereal containing malt. Later the child eats dessert at lunch of a chocolate brownie containing malt and chocolate. Then let's say that the child has a chicken dinner, complete with gravy using a maltodextrin containing product, homemade cookies for dessert baked with flour containing malt, and salad with dressing containing vinegar. The child is a walking chemical repository. Now it's time for homework and bed. I see children in my office who eat like this, and whose parents complain that their children have tantrums over homework, won't go to bed, can't fall asleep, and are grumpy in the morning. Many such children then receive powerful psychoactive drugs and years of expensive therapy, when really they could do fine if they ate better.

We need to understand that toxic chemicals work together to produce a greater effect.

How does the body handle the whole burden of toxic chemicals?

Since plants make toxic chemicals and humans eat plants, we must be able to detoxify toxic chemicals. These systems are studied because all the medicines which are given today are also detoxified by the same systems. A limited detoxification system also exists for the alcohol found in alcoholic beverages. Nowadays, the detoxification system is also asked to handle chemicals such as pesticides found in food and industrial chemicals inhaled in the work place.

What do we know about such systems?

Toxic chemicals work together to produce worse effects

For purposes of academic study, toxic chemicals usually are studied as single entities. For example, a work place chemical is given to an experimental animal and the effects are observed. But as we can see from the above lists of toxic chemicals, usually more than one or many toxic chemicals are coming in at once.

From studies on how drugs clear the body, we know that when there are two toxic chemicals, each is cleared less quickly than if either one were present alone. Think of the body as an auditorium partially filled with college students. If all have to leave at once, there will be lines at all the exit doors. Now if we put in a large number of high school students besides the college students, and if all have to leave at once, the lines to get out the doors will be longer. The students will be in the auditorium longer than if only one group of students was inside the auditorium.

The same is true of two sets of toxic chemicals instead of one set. The body can clear the toxic chemicals (get them out the exit doors) but the more toxic chemicals there are, the longer it will take to get all the toxic chemicals out. While the toxic chemicals are waiting to get out, they are doing whatever damage they do. The more toxic chemicals there are, the more damage each toxic chemical will do. If there are more toxic chemicals, the longer each toxic chemical will stay in the body, giving each chemical more time to do damage.

These additive effects of toxic chemicals mean that we do not know the effects of the total intake of food borne toxic chemicals. The more chemicals coming in, the more damage there will be, but we can't say how much. The university people cannot tell us because they do not study these questions. No government agency is concerned with this question.

Multiple low level chemical toxicities have not been studied

What is known about the impact of toxic chemicals on human beings? The majority of study in toxicology goes to the study of toxic chemicals used in the work place. There is much less study of chemicals occurring naturally. Many toxic chemicals are taken in at low levels over long periods of time.

Unfortunately there is almost no study of multiple chemical toxicities, especially when such chemicals are present at low levels and especially when such chemicals come in from the diet.

Usually when researchers study a subject, they want to keep things as limited in scope as possible, so to make for a cleaner study, they only study one toxic chemical at a time. At universities, professors get tenure based on the number of papers they publish. The incentives favor easily conducted studies rather than long term involved studies.

From studies of drug interactions, researchers know that if two drugs are given rather than one, that levels of both in the blood are likely to be higher than if only one was given. This is not a one plus one equals two correspondence. The body has a limited capacity to clear toxic chemicals. If two chemicals are present rather than one, the body clears each more slowly, and levels in the blood build up higher than if only one was present. In the case of yeast produced alcohols, the clearance rate is fixed. Ethyl alcohol can only be cleared so fast. The more alcohols are present, the longer the other, more toxic, alcohols will stay in the blood. The same is true for other toxic chemicals. In other words, more chemicals are likely to be more toxic than one chemical at a time.

Today toxic chemicals come from many sources. Chemicals are introduced during food processing either intentionally or unintentionally. There are chemicals of yeast and fungal origin. These chemicals come in from both the yeast present in the intestinal tract and from the diet. We don't know the impact of all the chemicals human beings take in.

The same problem of lack of study of multiple chemical toxicities applies to chemicals in foods also.

The big picture - food can be good and bad; what we know and don't know

As you can see, neither the general public nor doctors know much about what is in our food. I think it would be nice if we knew more about nerve poisons and other toxic chemicals in our food. Will more information be available in the future?

Let us look at how research is organized on food. I would love to go to a major university and ask for the "department of diet, food and health." That department would study and teach all about the food we eat, both positive and negative, and how it affects human health. Right now, scientists in departments of nutrition and biochemistry study the parts of food individually, the nutrients and energy that we need to survive and grow and for all the body's physiological and biochemical processes to work. But the study of diet, food and health, is much more than the study of nutrients.

Diet is both positive and negative

The study of diet and food is more than the study of nutrients. It includes both what is positive and negative. These negative parts of our diet include chemicals found naturally in food as well as those introduced during food handling and preparation.

I studied nutrition in graduate school for two years before glimpsing an understanding that food and diet are much more than nutrition. On a practice oral exam, a professor asked me to tell him about his cup of coffee. A cup of coffee contains water and caffeine and many other chemicals which are not nutrients. Caffeine is a very active chemical which affects significantly a number of body processes, but caffeine is in no sense a nutrient. Similarly, there are many other chemicals found in food which are very active and affect body processes, but which are not nutrients. The study of the whole diet is much more than the study of nutrition alone.

Let us look at an example from the field of cancer. Any time any food containing fat is cooked over a charcoal grill, cancer causing chemicals are formed which are called polycyclic aromatic hydrocarbons. These chemicals are very potent cancer causers. Based on the rather common use of charcoal grills in our society, there is relatively little knowledge about these chemicals among average people. Why is this? At least one reason is that no one in universities or even in government labs is specifically assigned or has taken the responsibility for understanding how these chemicals affect human health and cancer causation. These chemicals are a significant negative part of the diet. However, no one is assigned to study the negative parts of our diet. If there were a department of the diet, food and health, at universities, there would be such people.

Dietary toxicology: the study of bad chemicals in the diet

I term this field of studying the negative parts of the diet and food "dietary toxicology". Based on my study of what is known about chemicals in the diet, dietary toxicology is much more powerful at explaining why people do not feel well than is a field like nutrition. The negative parts of our diets affect us and cause health problems, because these chemicals are toxic. How diet affects health really is much more a study of what is negative or toxic in the diet, rather than the study of the quantity of nutrients present. As I showed you earlier, we eat numerous chemicals which affect our brains and nervous system. Unfortunately, this part of our diet receives little research attention.

I have already given the example of the numerous sedative chemicals in the diet. These chemicals affect the ability to concentrate and be productive. There is almost no knowledge available to the average person about such chemicals. Yet these chemicals significantly affect both adults and children. This book presents what I know and have been able to discover about these chemicals in food. What is presented here may be only a small part of what we should know. I only wish that there was more information available. Lists of chemicals found in food are hard to find. Worse yet, we don't know the ability of the body to handle toxic chemicals from foods when these chemicals are considered as a totality.

These chemicals should be studied as a totality

When one makes a list of the toxic chemicals that the body must handle, the list is long. No one has ever studied the toxicity of all these chemicals as a group. We need a full answer to the main question, "How and to what extent do these chemicals as a totality affect human health and disease?"

I wrote this book to show the therapeutic power of eliminating these chemicals from the diet. Awareness is the first step.

Eliminating such chemicals from the diet is helpful

Even though research on these chemicals is not anywhere near as complete as anyone of us would like, I can say that their removal helps people feel better. I will cite some disorders and cases where removal of such chemicals is beneficial in the next chapters. From reviewing such cases we can understand that these chemicals have a major impact on human health.

A word about pesticides

This book is not about pesticides, insecticides and herbicides. In general, these chemicals are more likely to be taken in from foods higher up on the food chain. Such foods are meat and fish. Vegetarians are much less likely to take in such chemicals. The World Wildlife Fund has made studies of such chemical intake.

Notes

[1]Lingappa, B.T., Prasad, M., Lingappa, Y., Hunt, D.F.., and K. Biemann. Phenethyl Alcohol and Tryptophol: Autoantibiotics Produced by Fungus Candida albicans. *Science.* 163:192-193, 1969.

Rankine, B.C. Formation of higher alcohols by wine yeasts. *Journal of the Science of Food and Agriculture.* 18:583-589, 1967.

Berry, DR., and D.C. Watson. Production of organoleptic compounds. In *Yeast Biotechnology,* Berry, DR., Russell, I and Stewart, GG., Eds., London, Allen and Unwin, 1987, p. 345.

[2] Gonzalez, A.M.T. and M. G. Chozas. Volatile components in Andalusian vinegars. *Z. Lebensm Unters Forsch.* 185:130-133, 1987.

[3] Barrett, J., Champagne, C.P., and J. Goulet. Development of bacterial contamination during production of yeast extracts. *Appl Environ Microbiol,* 65(7):32613, 1999.

[4]Barrett, J., Champagne, C.P., and J. Goulet. Development of bacterial contamination during production of yeast extracts. *Appl Environ Microbiol,* 65(7):32613, 1999.

[5] Gonzalez, A.M.T. and M. G. Chozas. Volatile components in Andalusian vinegars. *Z. Lebensm Unters Forsch.* 185:130-133, 1987.

[6] Croft, C. C. and L. A. Black Biochemical and morphologic methods for the isolation and identification of yeast-like fungi. *J. Lab. clin. Med.*, 23, 1248-58, 1938.

[7] ibid.

[8]Briggs, D. E., Hough, J. S., Stevens, R. and T. W. Young. *Malting and Brewing Science*, Volume 1 Malt and Sweet Wort, Chapman and Hall, New York, P.88-89, 1981.

[9]Muchmore, A. V., Decker, J. M., and R. M. Blaese. Spontaneous Cytotoxicity by Human Peripheral Blood Monocytes: Inhibition by Monosaccharides and Oligosaccharides. *Immunobiology*. 158:191-206, 1981.

Muchmore, A. V., Decker, J. M., and R. M. Blaese. Evidence that specific oligosaccharides block early events necessary for the expression of antigen-specific proliferation by Human lymphocytes. *The Journal of Immunology*. 125(3):1306-1311, 1980.

[10]Briggs, D. E., Hough, J. S., Stevens, R. and T. W. Young. Malting and Brewing Science, Volume 1 Malt and Sweet Wort, Chapman and Hall, New York, P.88-89, 1981.

[11] Wood, B.J.B. Progress in Soy Sauce and related fermentations, in *Progress in Industrial Microbiology*, ed. M. E. Buchell. Elsevier, Amsterdam, v. 19, pp. 373-410, 1984.

[12] Ostovar, K. and P. G. Keeney. Isolation and Characterization of Microorganisms involved in the fermentation of Trinidad's Cacao Beans. *Journal of Food Science*. 38(4):611-17, 1973.

[13]Jones, K.L. and S. E. Jones Fermentations involved in the production of cocoa, coffee and tea, in *Progress in Industrial Microbiology*, ed. M. E. Buchell. Elsevier, Amsterdam v. 19, pp. 411-456, 1984.

[14] Ibid.

[15]Riley, K. A. and D. H. Kleyn. Fundamental Principles of Vanilla?Vanilla Extract Processing and Methods of Detecting Adulteration in Vanilla Extracts. *Food Technology*. 43(10): 64-77, 1989.

[16]Cole, R.J. and Cox, R.H. *Handbook of Toxic Fungal Metabolites*. Academic Press, New York, 1981.

[17]Frazier, W. C. and Westhoff, D.C. *Food Microbiology*. McGraw-Hill Book Company. New York, 1978.

[18] Ibid.

[19] Ibid.

[20]Samonis, G., Anaissie, E.J. and G.P. Bodey. Effects of broad spectrum antimicrobial agents on yeast colonization of the gastrointestinal tracts of mice. *Antimicrobial Agents and Chemotherapy.* 34:2420-2422, 1990.

[21]Yeast make many alcohols, both ethyl alcohol and other much more poisonous alcohols such as 1-propanol, 2-propanol, 1-butanol and 2-butanol and phenyl ethyl alcohol (Rankine, 1967, Lingappa, et al, 1969, and Berry and Watson, 1987).

Lingappa, B.T., Prasad, M., Lingappa, Y., Hunt, D.F.., and K. Biemann. Phenethyl Alcohol and Tryptophol: Autoantibiotics Produced by Fungus Candida albicans. *Science.* 163:192-193, 1969.

Rankine, B.C. Formation of higher alcohols by wine yeasts. *Journal of the Science of Food and Agriculture.* 18:583-589, 1967.

Berry, DR., and D.C. Watson. Production of organoleptic compounds, in *Yeast Biotechnology.* Berry, DR., Russell, I and Stewart, GG., Eds., London, Allen and Unwin, 1987, p. 345.

[22]Yeast make many alcohols, both ethyl alcohol and other much more poisonous alcohols such as 1-propanol, 2-propanol, 1-butanol and 2butanol and phenyl ethyl alcohol (Rankine, 1967, Lingappa, et al, 1969 and Berry and Watson, 1987).

Lingappa, B.T., Prasad, M., Lingappa, Y., Hunt, D.F.., and K. Biemann. Phenethyl Alcohol and Tryptophol: Autoantibiotics Produced by Fungus Candida albicans. *Science.* 163:192-193, 1969.

Rankine, B.C. Formation of higher alcohols by wine yeasts. *Journal of the Science of Food and Agriculture.* 18:583-589, 1967.

Berry, DR., and D.C. Watson. Production of organoleptic compounds. In *Yeast Biotechnology.* Berry, DR., Russell, I and Stewart, GG., Eds., London, Allen and Unwin, 1987, p. 345.

[23]Browning, E., *Toxicity and Metabolism of Industrial Solvents.* Elsevier, NY, 1965.

[24] Browning, E., *Toxicity and Metabolism of Industrial Solvents*. Elsevier, NY, 1965.

Phenyl ethyl alcohol causes coma. Large amounts of vanilla cause coma in Jenner, D. M., Hagan, E. C., Taylor, J. M., Cook, E.L. and O. G. Fitzhugh. Food Flavourings and Compounds of Related Structures. I. Acute Oral toxicity. *Food and Cosmetics Toxicology*. 2:327-343, 1964.

[25] Berry, DR., and D.C. Watson. Production of organoleptic compounds, in *Yeast Biotechnology*, Berry, DR., Russell, I and Stewart, GG., Eds., London, Allen and Unwin, 1987, p. 345.

[26] Acetone causes gastrointestinal irritation, narcosis, injury of the kidney and liver and can cause coma. W. B. Deichmann and H. W. Gerarde, Academic Press, New York, 1969, p. 64.

[27] Torner, M. J. Martinez-Anaya, M. A., Antuna, B., Benedito de Barber, C. Headspace flavour compounds produced by yeasts and lactobacilli during fermentation of preferments and bread doughs. *Int. J. Food Microbiol*. 15(1-2):145-52, 1992.

[28] This reference is found in RTECS, the registry of toxic chemicals, and is in Russian, so it is not listed here.

[29] Mitochondria are small structures found in cells which make the body's energy. Yeast in addition produce the mitochondrial poison H_2S, hydrogen sulfide (Rankine, 1964, and Berry and Watson, 1987). H_2S is a significant brain depressant (Gosselin, et al, 1976).

Rankine, B.C. Hydrogen Sulphide production by yeasts. *Journal of the Science of Food and Agriculture*. 15:872-877, 1964.

Berry, DR., and D.C. Watson. Production of organoleptic compounds, in *Yeast Biotechnology*, Berry, DR., Russell, I and Stewart, GG., Eds., London, Allen and Unwin, 1987, p. 345.

Gosselin, RE., Hodge, H.C.., Smith R.P. and MN Gleason. *Clinical Toxicology of Commercial Products*. 4th edition, sect. III, PP.169-173 and 271-274.

[30] George, WL., Rolfe, RD. and Finegold, S.M.. Clostridium difficile and its cytotoxin in feces of patients with anti-microbial agent associated diarrhea and miscellaneous condition. *J. Clin. Microbiol*. 15:1049, 1982.

[31]Clostridium synthesizes p-cresol and other Clostridium species synthesize phenol, (Elsden, et al., 1976). Phenol is used as an antiseptic and kills bacteria. Both phenol and p-cresol are toxic CNS depressants (Gosselin, et al, 1976, and Deichmann and Keplinger, 1981). The effects of phenol and p-cresol would be additive with the yeast alcohols and H_2S.

Elsden, S., Hilton, MG and JM Waller. The end products of the metabolism of aromatic amino acids by Clostridia. *Archives of Microbiology.* 107:283-8, 1976.

Deichmann, W. B., and Keplinger, ML. Phenols and phenolic compounds, in *Patty's Industrial Hygiene and Toxicology,* V.2A, John Wiley and Sons, New York, 1981, c.36, pp.2597-2601.

Gosselin, RE., Hodge, H.C.., Smith R.P. and MN Gleason. *Clinical Toxicology of Commercial Products.* 4th edition, sect. III, PP.169-173 and 271-274.

[32] Gonzalez, A.M.T. and M. G. Chozas. Volatile components in Andalusian vinegars. *Z. Lebensm Unters Forsch.* 185:130-133, 1987.

[33]This reference is found in RTECS, the registry of toxic chemicals, and is in Russian, so it is not listed here.

[34]Hough, J. S., Briggs, D. E., Stevens, R. and T. W. Young. Brewery Fermentations. *Malting and Brewing Science*, 2nd ed., Chapman and Hall, London, vol. 2, pp. 462-471, 1982.

[35]Nishie, K., Waiss, A. C. Jr., and A. C. Keyl. Pharmacology of Alkyl and Hydroxyalkylpyrazines. *Toxicology and Applied Pharmacology.* 17:244-249, 1970.

[36]Liardon, R. and G. Philippossian, *Z. Fur Lebensmittel-Untersuchung and Forschung*, 167 (1985), 180-85.

Liardon, R., and S. Ledermann. *Z. Fur Lebensmittel-Untersuchung and Forschung.* 170(1980):208-13.

[37] Zak, D. L., Ostovar, K. and V. G. Keeney. Implication of Bacillus subtilis in the synthesis of tetramethylpyrazine during fermentation of cocoa beans. *J. Food Sci.* 37(1972), 967-68.

[38] Deck, R.E., and Chang, S. S. Identification of 2,5-dimethylpyrazine in the volatile flavour compounds of potato chips. *Chem. Ind.* No. 30, 1343-1344, 1965.

[39] Mason, M. E., Johnson, B., and Hamming, M. Flavor components of roasted peanuts. Some low molecular weight pyrazines and pyrrole. *J. Agr. Food Chem.* 14, 454-460, 1960.

[40]In addition to the pyrazines, soy sauce contains 4 ethyl guaiacol, methyl hydroxy furanones, alkyl pyrazines, alkyl pyridines and amides (8) (Liardon and Philippossian, 1978, Liardon and Ledermann, 1980). The alkyl pyrazines are highly sedative. In larger doses they can be lethal. Pyridines are industrial solvents and they cause headaches, nervousness, dizziness and insomnia, nausea and anorexia, frequent urination, dermatitis, liver and kidney damage.

Liardon, R. and G. Philippossian, *Z. Fur Lebensmittel-Untersuchung and Forschung*, 167 (1985), 180-85.

Liardon, R., and S. Ledermann. *Z. Fur Lebensmittel-Untersuchung and Forschung.* 170(1980):208-13.

[41]Law, B. A. Microorganisms and their enzymes in the maturation of cheeses. in *Progress in Industrial Microbiology*, ed. M. E. Buchell. Elsevier, Amsterdam v. 19, pp. 245-284, 1984.

[42]Cole, R.J. and Cox, R.H. *Handbook of Toxic Fungal Metabolites.* Academic Press, New York, 1981.

PART II
How Yeast Chemicals Affect Our Health

Chapter 3

Yeast Chemicals and Your Health

- *Abdominal Pain*
- *Chronic Constipation*
- *Excessive Gas (Flatulence)*
- *Headaches and Migraines*
- *Eating Disorders and Anorexia nervosa*
- *Chemical Sensitivity*
- *Chronic Fatigue Syndrome*
- *Tourette's Syndrome*
- *Seizures*

I have found that treating the intestinal yeast Candida albicans helps with many medical problems that appear to be unrelated. The question is, why would one type of treatment work on so many different problems that conventional medicine would treat by sending people to many different specialists?

When we examine how the yeast Candida albicans works in our bodies, we will think about medical problems differently. Rather than a particular problem being a particular illness, with its own name and course of treatment, we can think of a particular problem as a set of symptoms. For example, chronic constipation usually is treated as a problem in itself; however, it really is a symptom that the intestine is not functioning properly. So is abdominal pain.

I have found that there are classes of symptoms that are related by how yeast works in the body. We can understand these classifications if we look at medical problems which yeast chemicals cause, and other medical problems caused by yeast interacting with and fooling the immune system. In this chapter, I will discuss part of the first group of medical problems, those which yeast chemicals cause. These problems are: chronic constipation, abdominal pain, headaches, the eating disorder anorexia nervosa, chemical sensitivity, chronic fatigue syndrome, Tourette's Syndrome and seizures. Other medical problems which yeast chemicals cause, but which are slightly different, are discussed in separate chapters because they require more complex explanation.

The disorders of chronic constipation, abdominal pain, headaches, anorexia nervosa, chemical sensitivity, chronic fatigue syndrome, seizures and Tourette's disorder are complicated. The traditional medical world does not completely understand them. Most of the patients I see with these disorders have tried numerous conventional treatments, none of which have worked. Their doctors have told them either the problem does not exist ("it's in your head"), that stress is causing the problem, or that they should "just learn to live with it." In all of these disorders, the body's nerves or brain or both do not work properly. No one knows why the nerves and brain do not work as well as they should.

My clinical observations have shown me, and can show you, that treating Candida albicans improves these conditions significantly. Recall my discussion from Chapter 2 about how the yeast Candida makes many chemicals which are toxic to the nerves and brain. These chemicals may be part of the reason why nerves and brain do not function as well as they should in people with these disorders. I will give in this chapter specific cases and my best explanations on how yeast chemicals play a role in these disorders. I will show that this anti-yeast treatment works.

Chronic Constipation

Chronic constipation improves significantly with anti-yeast treatment. To understand why, we need to understand some facts about the gut. The intestinal tract has its own nervous system and brain. A large system of nerve cells controls the movement of the intestine. The food moves along at exactly the right rate, mixing with digestive enzymes and fluids.

Yeast chemicals are toxic and slow down the nerve cells in the gut. If these nerve cells are sedated, they cannot do their job properly of controlling the movement and activity of the gut. The intestinal wall may not move along at the right rate. Slowing the gut nerve cells appears to lead to constipation. When a person who suffers from chronic constipation follows anti-yeast treatment, the sedating chemicals no longer slow down the nerve cells. The gut works better, so the patient is no longer constipated. Unlike treatment with laxatives which only treats the symptoms, anti-yeast treatment targets the cause of the constipation so that the problem does not return, provided the patient stays on the anti-yeast diet system.

Abdominal Pain, Chronic Constipation, and Flatulence

Abdominal pain usually improves when the patient follows anti-yeast treatment. Let us look at how yeast might cause such pain. Yeast adhere to the inner intestinal wall. Yeast also makes chemicals. A number of these chemicals are listed in the registry of toxic substances as skin and eye irritants, and one is listed as being very destructive of human tissue.[1] These chemicals will irritate the inner intestinal wall. If a person eats foods such as malt, which provides growth factors for the yeast, the yeast will become more active and make more chemicals. Some of these chemicals will irritate the intestinal wall, causing pain. If the yeast is treated, the chemicals it makes will no longer be there to irritate the intestinal wall.

Cases - Abdominal Pain and Chronic Constipation

Kim

Kim, 37, came to see me because she was having "cranky spells," depression, and bad bouts of vaginal yeast infections after antibiotics. She had breath and foot odor.   She had had asthma attacks during her pregnancies.   An allergist found she had high IgE (the antibody of allergy). She was allergic to mold and felt like she was choking. When the spring came, she would get hives and her fingers would swell. She also had abdominal pain. She told me that eating vinegar caused her nose to fill up, and that some wines and aged cheese bothered her. She had stopped the birth control pill about two years previously. Afterwards, her IgE level decreased. Kim had occasional heartburn, headaches, abdominal bloating and finger swelling after she ate certain foods.

Kim started the 4 Stages diet and nystatin. She returned four weeks later. Kim was taking one half teaspoon of nystatin four times per day. She felt great. Kim's abdominal bloating was gone and she had less finger swelling. She had no headaches or stomach aches and her nose was clear. Her sore throat had gone away. She had not had any colds. This was remarkable because this appointment was in the winter time when Kim usually caught colds. Kim's chronic constipation had cleared. Prior to starting my prescribed program, Kim could only have a bowel movement every ten days. After four weeks on the diet and taking nystatin, Kim had regular bowel movements.

Kim's only complaint was that her period felt like it was coming early and her breasts were tender.

I would comment that in patient care, patients commonly find that lesser problems bother them more after the major problem clears. In time, the lesser problems also usually clear up.

Alecia

Alecia came in at the age of six for problems with anger and abdominal pain. She would become angry when she did not get what she wanted. She would threaten to destroy things and scream. She told her mother that her anger would build and that she could not help herself. She also had trouble falling asleep. She would wake during the night. Alecia had abdominal pain during the day, and now such pains were occurring at night. Her daily abdominal pain had started the year before when she was only five years old. Alecia also had other pains also. She feared failure. Alecia had a history of many ear infections. She had tried a milk elimination diet. She had had a major seizure at eight months and had taken an anti-seizure drug for one year afterwards.

I prescribed for Alecia the 4 Stages diet and nystatin. Alecia came back six weeks later. She was following the 4 Stages diet and taking one quarter teaspoon of nystatin four times a day. Alecia's mother reported that Alecia was much better. She had had only two angry outbursts since starting the program, one of which occurred after eating regular bread. Alecia was now sleeping well and she had much less abdominal pain.

Case of flatulence (intestinal gas)

Deangelo

Deangelo, 44, told me he had bad intestinal gas. He said that he expelled gas especially when exercising and at other times. Deangelo's problem with intestinal gas had been bad for a year. He had had several tests which were all normal. Deangelo noticed that his problem with intestinal gas started right after taking the antibiotic tetracycline to treat strep throat. He had had tetracycline several times in the past. He also complained of generalized itching when he was tired. He was able to alleviate this with liquid garlic. A heart catheterization had shown evidence of a mild past heart attack. Deangelo had stopped smoking three years previously, stopped drinking alcohol within the past year and had started exercising. He had already tried changing his diet by increasing his intake of whole grains, peanut butter, chicken and pork, and eliminating beef and sweets. He was taking alfalfa and watercress concentrate and beta carotene. Beef and stress both made the gas worse.

Deangelo started taking one fourth teaspoon of nystatin four times per day and began following the 4 Stages diet. After three weeks, Deangelo came back. He reported that after one week, he had experienced tremendous results. The gas had cleared entirely. A testicle which had been enlarged for some time (not mentioned at the first visit) had flared up but was now reduced in size. Deangelo said that he noticed an increase in his energy level, even though at the first visit he was not aware that he was fatigued. His itching was better, but it still bothered him once in a while.

Treating intestinal yeast improves abdominal problems

From these cases, we can see that treating Candida albicans greatly improves the lives of people who suffer from abdominal pain, excessive flatulence and chronic constipation. These problems usually clear up within four to six weeks and, as long as the patient steers clear of the wrong foods, the problems don't return.

Headaches and Migraines

Nobody knows what exactly what causes headaches. Many people suffering from migraines end up taking medications to prevent the onset of the debilitating pain that migraines cause. Changing food choices to exclude foods such as vinegar, malt, chocolate and pickles, reduces headache frequency and severity. Many other health practitioners and I know this from treating patients who suffer from headaches. Other health care practitioners also provide lists of foods to avoid. Although not completely the same as the list of foods to avoid on the 4 Stages Diet, many of the foods overlap. These foods include peanuts, peanut butter, chocolate, bananas, aged cheese, and others. The only way we could explain these clinical results would be to examine whether these foods contain something which causes headaches. What might these foods contain which causes headaches?

Some of these foods contain the chemical tyramine. Medical researchers have studied tyramine. Tyramine is found in foods which show up on other lists of headache causing foods such as cheese and red wine. As I discussed earlier, malt contains tyramine. This is one reason why eliminating malt from your diet decreases headache frequency.

Tyramine is only one chemical of many. Any chemical which can affect the brain and the blood vessels next to the brain could be a possible headache causer.

In Chapter 2, I explained that the yeast Candida albicans makes many chemicals which are toxic to the brain. These chemicals include toxic alcohols, as well as the powerful brain poison hydrogen sulfide. Vinegar contains the chemical ethyl acetate, which slows the brain. Vinegar also contains the toxic alcohol methanol. Malt contains twenty chemicals which slow down the brain.

Fungi also make chemicals which are toxic to the brain. I have found from treating many patients that foods contaminated by yeast or fungus, for example, peanuts, or foods in which yeast has been present, such as vinegar, all can cause headaches.

Toxic chemicals enter the body in food and from the intestinal yeast. The liver should clear most of these chemicals but it does not clear all. No one knows exactly what these chemicals do when they get to the blood vessels of the brain. Why might these chemicals play some role in causing headaches?

In the previous section on abdominal pain, I noted that the registry of toxic chemicals lists some of the yeast chemicals as skin and eye irritants. These chemicals can circulate and irritate the inner linings of the blood vessels of the brain. Whether the liver clears them out enough and whether the circulatory system dilutes these chemicals enough to reduce their irritant quality, I do not know. They may directly irritate the blood vessels of the brain, causing pain.

A second and not mutually exclusive explanation of how these chemicals work is to think about how the blood vessels might react to toxic chemicals. Conceivably the chemicals could cause the cells lining the blood vessels to produce inflammation and swelling. This would prevent such chemicals from passing through the cells of the blood vessels and into the brain. This swelling and inflammation would keep the chemicals in the blood rather than allowing them to enter the brain. The result would be pain, but the brain would be protected. The pain, or headaches would be an effect of the body's response to toxic chemicals.

We don't really know for sure what the mechanism is that causes these headaches. I have proposed two hypotheses, based on my observations that anti-yeast treatment decreases headaches. The best solution for headaches is to avoid foods containing such toxic chemicals that could irritate blood vessels or cause an inflammatory reaction. These foods are vinegar, malt, chocolate, pickled foods alcoholic beverages, nonalcoholic beer, peanuts, soy sauce, worcestershire sauce, and cottonseed oil. These are the foods eliminated in Stage I of the 4 Stages Diet.

Just eliminating these foods dramatically reduces the frequency and severity of headaches.

Cases of headaches

Norma

Norma, 37, told me she felt emotional and cried easily, worse before her menstrual period. These symptoms began prior to her marriage 16 years earlier when she started the oral contraceptive pill, which she took for nine months. She then had two children. She had headaches every day, but a tooth retainer had reduced the headaches to 1 to 2 per week. She had regular menstrual cycles with heavy bleeding. She had received numerous antibiotics in the past, more frequently since being on the pill.

After seeing me, she started the 4 Stages diet and began taking one half teaspoon of nystatin four times per day.

She came back three weeks later and said that her headaches were gone. Her last period was better. She could deal with work better.

She came back after seven weeks of treatment. Norma reported that everything was fine, but she said that if she missed a few nystatin doses she started to feel bad.

Carol

Carol, 43, came to see me because she had suffered from premenstrual syndrome, extreme headaches, a plugged up nose and breathing troubles. This had gone on for the past 6 years. Allergy testing showed a few fall allergies. The sinus problems were so bad she had had sinus surgery the year before. When Carol came to me she was getting nosebleeds for a week before her period. Her nose was swelling up, especially at the incision sites. She still had headaches. She had a hiatal hernia, so she had been told to avoid coffee and chocolate before her period. Carol had intense chocolate craving during this time. She had a stressful job and could not concentrate. Concentration was worse before her period. She also had bloating and gas premenstrually. She had patches of dry skin on her scalp and face. Her skin itched before her period. She had no nasal congestion or headaches at other times of the month. Her periods were only 21 -28 days apart, so she experienced these symptoms every few weeks. Carol had taken many antibiotics the year before the problem started. She had three children.

After discussing her problems and symptoms, I prescribed the 4 Stages diet for Carol. Two and a half months after being on the diet and taking nystatin, Carol came back feeling "pretty good". Her headaches had decreased from being debilitating for days at a time, to only the day before her period. They were much less intense. She had more energy. Her nose was still plugging up and bled some, but less. The itching on her head was gone. The dry patches of skin on her ears and eyebrows were gone but the dry patch on her scalp was still present. Her bloating was gone. Besides all these physical symptoms improving, she craved chocolate much less.

The 4 Stages diet reduces headaches

Carol's and Norma's cases are but two of many examples of how changing one's diet to the 4 Stages Diet and taking nystatin reduces or eliminates severe headaches.

Anorexia nervosa and eating disorders

Anorexia nervosa is an illness in which people do not want to eat. They lose an unhealthy amount of weight and believe themselves to be fat. They reduce their food intake and may exercise a great deal. Women stop menstruating. Many patients die.

Medical scientists understand little about the cause of this disorder. Anorexia is treated as a psychiatric disorder, caused by psychological problems rather than physical problems. The treatment is to force feed people back to a normal weight. Often, psychoactive drugs are prescribed. Doctors hope that the distortions in thinking will improve when there are more calories coming in. However, many patients begin to starve themselves again after they are refed.

There is at least some evidence that anorexia nervosa is caused by the Candida albicans. Let me explain.

When I was learning child psychiatry, I took care of a number of anorexic patients. Most did not fight the refeeding, but one did. She was twelve years old and weighed 67 pounds. It took six of us to hold her down and one to insert the feeding tube. She had to be restrained to prevent her from pulling out the tube. I could not imagine what could push someone so hard to not eat when most of us enjoy eating.

Months later I connected some thoughts and ideas. I had been studying toxic yeast chemicals. Such chemicals harm the brain. The brain does not want these chemicals coming in. How can it stop them from coming into the bloodstream? One way would be to not take in any foods containing such chemicals. This solution may be helpful, but what if the main source of yeast chemicals is the internal intestinal yeast? How could the brain reduce the chemical load coming in from intestinal yeast? One way would be not to eat. The yeast need food to be active. Without food, they slow down. The brain could reduce the load of yeast chemicals coming in by causing the person not to eat. Not eating would protect the brain from yeast chemicals.

The person in whom this is happening may not have any conscious understanding of what is happening. The person explains this to the people around him or her as "I am fat and therefore I should not eat." In reality, the brain is protecting itself from toxic yeast chemicals by reducing food intake or even starving.

This sequence of events is hard to prove or disprove. I do not know if this sequence occurs exactly as I have suggested. Still, patients with anorexia do better with anti-yeast treatment (see cases below).

Foods such as vinegar and malt contain yeast chemicals and other chemicals toxic to the brain. The most concentrated sources of such chemicals in the diet are in vinegar and malt. Vinegar is found in many sauces and malt is baked into many baked goods. Feeding foods containing yeast chemicals such as vinegar and malt to anorexic patients is very unfortunate. They may gain weight, but their brains continue to tell them that bad chemicals are coming in. They say to others that they are fat. I think that this means these yeast chemicals are too much for them. Anorexic patients may fight this refeeding. Why? Because these chemicals are bad for the brain. Nothing changes. The intestinal yeast chemicals and dietary chemicals are still there. As soon as the forced refeeding is over, the anorexic patients go back to starving themselves.

Anorexia is a severe, life-threatening illness. Rather than simply force feeding any foods at all, patients should be offered foods from Stage IV of the 4 Stages diet (those which

have no yeast, mold, or fermentation), and should be given
nystatin. This treatment can work and has no serious side
effects.

Anorexia Cases

Dr. Orian Truss reported one case of cured anorexia by
treating yeast and using nystatin in his book *The Missing
Diagnosis*. In this case a hospitalized anorexic woman near
death received nystatin because the doctor on call saw that she
had oral thrush (Candida in the mouth). The woman made a
full recovery in a few days.

The following case is from my clinical experience.

Dee

Dee, 40, complained of not feeling good about eating.
She lacked energy and had high cholesterol. She was
depressed and did not think that she could expend the
energy to change her diet. She was afraid that if she ate
more, she would become fat and unattractive. She
could not handle the low cholesterol diet her doctor had
recommended. She had terrible mood swings. She
was smoking and drinking a lot of coffee. Dee had had
bouts of severe depression. She first became depressed
about 8 years previously while going through a divorce.
Six years later, Dee was hospitalized for depression.
Dee had again become depressed shortly before coming
to see me. She had been labeled as having an eating
disorder, and she panicked over eating three meals a
day. Her throat would feel as though it would close
when she thought about eating so much. She was
eating a little bit, but she acknowledged that she was
killing herself by not eating and by smoking. In the
past, her husband had pushed her to eat and she became
upset at these times. She had been on a number of
medications. She complained now about abdominal

pain after eating, which depressed her and made her feel like not eating anything. This patient also had a history of drinking and using drugs, but was sober now. Dee appeared thin and looked older than her age.

Dee came back four months later and stated that she followed the diet and took nystatin and did better. Her appetite returned; she had eaten much better and her energy and mood had improved. She looked much better and had gained some weight. She had run out of nystatin. I suggested that she continue the diet and nystatin.

Eight months after starting treatment, Dee said that she was doing fine. She was attending college, was able to concentrate and her grades were good. If she did not eat her mood and energy decreased, but eating caused pain and distention of her belly. She was eating twice a day.

Unfortunately, Dee later only complied partially with the treatment. She did not follow the diet completely, and did not always take nystatin.   However, she improved to the point where she could function even with less than total adherence to the treatment, and Dee maintained those gains.

Eating disorders can be debilitating. Changing the foods one eats, or in the case of anorexia nervosa, refeeding with yeast-free foods, can have significant results.

Chemical Sensitivity

Some people feel fatigued, have headaches or other symptoms from inhaling chemicals such as perfumes, diesel fumes or paint. These patients may improve if treated for intestinal Candida.

The problem of yeast toxic chemicals causes chemical sensitivity. We know from previous discussion that yeast makes toxic chemicals. These yeast chemicals can circulate in the body. Such chemicals are similar in structure to some industrial solvents. One yeast chemical, ethyl acetate, is found in glue thinner. I suspect that in some way, the yeast chemicals and any chemicals coming in by breathing are additive in their effects on the body and on the brain. Because the burden of such chemicals in the body is already high due to yeast in the body, adding such chemicals by inhaling air containing these or other chemicals may increase the overall effects of all of these chemicals. At some point, everyone's body has a saturation point for tolerating a chemical load. When the chemical load passes saturation, people experience symptoms of chemical sensitivity. When yeast chemicals in the body are reduced, airborne chemicals become easier to tolerate.

Chemical Sensitivity Case

Tom

Tom, 31, came to see me because he had severe chemical sensitivities and chronic fatigue. Breathing shampoos, perfumes and diesel fumes caused pain and tightness in his chest and a funny sensation in his head. He then would get a cold which would go to his chest. Tom needed an antihistamine and allergy shots to help him get through the cold. He had had Strep throat two times in the past, about ten years previously. Later, after a car accident, he began to feel worn out and lazy. He then started to get Strep throat followed by antibiotic treatment, frequently. He could hardly work. He had surgery to remove his tonsils and he had nasal surgery for a deviated septum. He started allergy shots and felt better. However, when he went back to work in a machine shop, his nose would clog. He had tried going to Arizona to live. Hopefully, the climate would be easier, but the same thing happened. Steroid shots helped, but the problems persisted in the machine shop.

Tom knew about this anti-yeast diet from another patient of mine. He tried the 4 Stages diet on his own; this helped. Tom used to drink beer, but beer gave him headaches. He had quit two years previously. When he returned to Wisconsin, he developed asthma. He went back on allergy shots for molds; this helped. Tom was now working in an office. He was still taking shots. Tom said that he became sick if he went off the 4 Stages diet. His energy level was still weak. His ears were stuffed at times and he had occasional headaches.

When Tom came to see me, he had already started the diet. I prescribed nystatin. Within six weeks, Tom's energy level had improved and his other symptoms were intermittent instead of constant.

After six months of treatment, Tom was taking three eighths teaspoon of nystatin four times a day and following the diet. Tom reported doing well, but still had intermittent nasal congestion. He had occasional chest pains. He felt that he had improved 70% on odors and that only one cologne was still bothering him and causing his chest pain. Diesel fumes no longer bothered him. His energy level was good.

Tom's case shows how following the 4 Stages diet and taking nystatin to kill intestinal yeast can improve chemical sensitivity.

Chronic Fatigue Syndrome

Chronic fatigue syndrome is a disabling condition which robs its sufferers of their energy to function in life. This syndrome has no known cause and no standard treatment. I have helped many sufferers of chronic fatigue syndrome by treating the yeast Candida albicans. Let me explain why I think that this treatment is helpful.

As I explained in Chapter 2, intestinal yeast makes chemicals which slow down the brain. These compounds are quite toxic (poisonous) to the nervous system. These compounds include toxic alcohols and acetone, as well as the powerful nervous system poison hydrogen sulfide.

Alcohols depress the nervous system. Acetone causes coma. These chemicals together with hydrogen sulfide can be viewed as putting the brain to sleep. These chemicals all slow down the brain. Then the brain may no longer work correctly. The liver should clear these chemicals so that they never reach the brain. However, in some people, the liver apparently does not clear these chemicals and they reach the brain.

In addition to having yeast that lives in their body, a fatigued person also gets yeast-type chemicals from foods which contain chemicals that also slow down the brain. I explained how this works in Chapter 2. Remember that vinegar and malt, the raw material used for making beer as well as many other foods, contain chemicals which slow down the brain.

Based on my clinical observations, these chemicals cause the brain and body not to be able to function at their normal level of energy. When the patient removes these chemicals, both by treating the intestinal yeast and by changing the diet to exclude chemicals which slow down the brain, the patient's energy improves.

Treating the yeast and changing the diet diminishes the burden of sedative chemicals. I have found that energy returns in patients who follow my treatment plan.

All the fatigued patients I have seen who have followed my 4 Stages diet and taken nystatin have gotten at least part of their energy back. Some of these people had been fatigued for years or had not even been able to get out of bed. Even these patients have had at least partial return of energy.

Cases of severe chronic fatigue

Nancy

Nancy, 47, told me she had "fibromyalgia." She was barely able to get out of bed. Nancy could go for days without sleeping even though she stayed in bed. She had leg twitching, pain and weakness. She could not walk. She had had headaches since the age of 18, and neck pain since age 21, leg weakness for 14 years and skin rashes since age 18 (seborrhea). Nancy had not been able to get out of bed for two years. Nancy had had eye twitches for a year.   Nancy had hip pain and she had had hives for six months.   Nancy was depressed, could not work, and had lost her friends and her marriage. She had tried acupuncture, vitamins, fasting and juicing. A colonic had helped for several months. A stool sample had showed Candida.

Nancy had taken many antibiotics as a child, which would have increased her intestinal Candida.   She also had vaginal bleeding for which she took oral contraceptive pills. She had been pregnant two times.

I started Nancy on the 4 Stages diet and nystatin.

Nancy came back four weeks later. She had improved significantly. Nancy reported that she had lost weight (she was not overweight to begin with.) Her leg twitching had stopped. Most importantly, Nancy was out of bed eight to nine hours per day. She continued to have leg weakness and pain, but she was sleeping less.

When she slept, she slept better. Nancy also tried some other kind of neck treatment, a neck injection. Her headaches were gone. Her overall body fatigue was not as bad. She had no neck pain and her legs were not giving out. She was taking nystatin at one quarter teaspoon four times per day. In other words, a bedridden woman got out of bed in four weeks, after starting the 4 Stages diet and nystatin.

Nancy came back to see me several years later. Her energy level had stayed improved for several years but she had relapsed after severe injuries in a car accident two years previously. Before this accident, she had remarried and had started to work, but the accident had stopped her. Treatment was restarted.

Edna

Edna, 48, was bothered by fatigue. She needed one to two naps per day and was sleeping ten to twelve hours per night. The rest of the time she felt sick, as if she were on the verge of getting a cold. She was also nauseated. She was disabled and was receiving social security benefits. She had quit work eight years previously at a college level job. Now she only worked very part-time. She had experienced fatigue for twenty years.

Edna had had low grade fevers regularly for three years. She used to have sore throats and headaches, then fevers, all with fatigue. She felt the worst in the afternoon. Her fevers were usually in the afternoon.

She had seen another doctor a few months previously, who had prescribed Diflucan (an antifungal drug), Valtrex (an antiviral drug) and gamma globulin injections (gamma globulin is the fraction of the blood which had antibodies for fighting some kinds of infections). She felt a little better.

She also had diarrhea and was told she had colitis. She had a colonoscopy in the past.

Edna stated that she ate a good diet. She had started part of the 4 Stages diet already, excluding vinegar and malt about six weeks previously and said she felt a little better. But she said many things had helped temporarily and she was afraid that these dietary changes would stop helping.

Edna was depressed over being sick. She had felt a little better in the last few weeks but the diarrhea was intermittent. She had a history of abdominal pain, although this did not bother her now.

She had mononucleosis fourteen years previously, which had lasted a long time. She had taken antibiotics as a child.

I recommended that Edna stay with the 4 Stages diet, making sure she was doing the entire diet, and start nystatin.

She came back five weeks later. Edna stated that she was feeling better. Her fevers were now only occasional instead of daily. She no longer needed to take naps. She was sleeping through the night for eight to nine hours. The sick feelings were all gone. Things did not bother her as much and she could handle stress better. She had much more energy.

Her diarrhea was gone. Her bowel movements were normal. Her mood was much improved. She no longer had abdominal cramping.

She stated she could now go back to work.

Twenty years of fatigue was over in eight weeks.

Tourette's Syndrome

Patients with Tourette's Syndrome suffer from both motor and vocal tics. Motor tics are repetitive motions of muscles which only are partially under voluntary control and may range from eye blinking to complicated motions of the trunk and arms. Vocal tics are the involuntary repetitive saying of words or short phrases, which can unfortunately include swear words. When both of these occur in the same person, the disorder is called Tourette's Syndrome or Tourette's Disorder. There is no known cure and the medicines used are "heavies" (antipsychotic medicines which have many side effects).

I have successfully treated several cases of Tourette's using anti-yeast therapy and nystatin. Why might this therapy be helpful? I offer the following thoughts. For a tic to occur, a center in the brain must fire, triggering the muscles to move. In normal individuals, such centers do not fire involuntarily. The reason that these centers do not fire involuntarily is that most of the brain is devoted to keeping the brain centers ready to work, but not actually working. The brain allows the part we want to be active to focus on what we want. For example, if I wish to reach out and pick something off a table, my arm and hand move to pick it up. This motion does not take my whole brain to do. The rest of my brain is making sure that I focus only on what I want to do. At the same time, the brain is making sure that, for example, my legs do not move when I reach out with my arm. The majority of the brain is inhibitory; that is, it keeps most of the brain ready but not actually working unless that part of the brain is needed.

In Tourette's, the inhibitory function appears to be decreased, so a brain center, instead of being told to wait until needed, can simply fire and do what it does, such as blink an eye. The problem is partially that a center is too active, but it is also that the rest of the brain is not working properly. The rest of the brain should be inhibiting this overactive center but does not.

We can understand how this might be the case if there are toxic sedative chemicals slowing down the brain. The parts of the brain which keep other centers ready but not actually working are themselves slowed down. Then if another center wishes to fire, it is not being inhibited as much as it should be and it fires.

Clinically treatment of intestinal yeast leads to fewer tics. The brain appears to function better. Then Tourette's symptoms diminish considerably. Let us look at the cases.

Cases of Tourette's Syndrome

Steve

Steve, 29, told me he had Tourette's Disorder and obsessive problems. He had tics in response to chewing sounds and other environmental stimuli. Hearing people lick their lips bothered him. He went over work constantly in his mind. He moved his right shoulder. He repeated words in his mind. He would touch things repeatedly. He did not feel comfortable around people. He was angry and wanted to withdraw from everyone. He was germ conscious, washed his hands often and did not like to go into public bathrooms. This problem had been worse in the past. He had occasional headaches. He was tired and had trouble getting up.

He had had tics since the age of 7. Earlier in his life his tics had been more physical. Now they were more mental. He can spend hours with thoughts now. He had been depressed and fatigued over his tics. Creative writing and drawing helped. So did being alone. Cutting down on sugar had helped. He said that he was hypoglycemic. He had had a bad reaction to a medication with "paralysis and shaking". He did not want any more regular medications. Steve had taken antibiotics in the past.

Steve had held a number of jobs. He had a bachelor's degree, but was not working. In my office, he spoke slowly and he was moving his right shoulder in tics. He kept his right hand covered and he seemed physically tense. His stepmother, also a patient, had encouraged him to come in.

Steve started the 4 Stages diet and nystatin. He came back a month later and stated that he did not think that his tics were better, but he was less fatigued and less depressed. His stepmother, however, reported that Steve's tics were less severe and less forceful. She said that he looked better to her, and I agreed. He was now looking for work. He was eating a lot of popcorn but was otherwise compliant with the diet. I saw only occasional movements. He appeared more relaxed and comfortable. I suggested that he continue the treatment since he was better in some ways.

Steve came back a month later and said that he was not sure that he was better. However, his stepmother and father said that he was better. He said that he still had obsessive thoughts and still felt depressed at times. However, he had been applying for jobs and was looking into going to graduate school. In my office, he smiled much more than when seen previously and he appeared more comfortable. He was moving around more quickly. I thought that he had improved at least some and suggested that he continue the treatment.

Within the next few months he called to say that he would not be coming to see me anymore because he was going out of state to graduate school.

This is a case in which the patient clearly had improved and was functioning well. His parents and I both saw visible improvement. The patient, however, did not want to believe he was better. Unfortunately this phenomenon occurs more than we like to believe and interferes with treatment. I discuss this in another chapter.

A case of Tourette's and severe side effects to conventional medications

Don

Don came to me at the age of 11 to get help with Tourette's and the side effects of medications. He had had Tourette's symptoms for four years and had been diagnosed five months previously. At that time, the tics had suddenly become worse with coprolalia (involuntary utterance of obscene words). His coprolalia was so bad that he was saying four letter words continuously from the time he got out of bed. Before that he had only facial tics. Before Don's tics got worse, he had not been moody. He was popular and was voted to student council. He was outgoing and was a smart kid. Now he was moody. He also had a symptom of obsessive compulsive disorder that he needed to confess things, and he had to come to his mother for this. Don also had a history of using antibiotics in the past.

Don's neurological examination was normal except for the tics. A number of antipsychotic medications were tried. Orap was tried for six weeks, but he gained twenty pounds. Then Risperdal (an antipsychotic) was tried but he had nausea and increased breast tissue. He lasted two months on Risperdal. Then Zyprexa (an antipsychotic) was tried which controlled the tics but he had scenes of visual rage. Paxil (an antidepressant) was added. The visual scenes decreased and were now intermittent. He was now fatigued, slept a lot, and was hard to rouse in the morning. He had mood swings.

When he was down or frustrated, his moods were extreme. He had gained twenty more pounds with the Zyprexa. Vitamins had been tried with no benefit. Craniosacral therapy had helped with mood swings.

By the time I saw Don, his tics were under better control. He displayed no coprolalia at the first visit. When he was first seen he was taking Zyprexa and Paxil (an antidepressant). He was having vocal and motor tics when seen and he had some thoughts of suicide. He was overweight and was seeing visual scenes of rage. He was 50 pounds overweight.

He started the 4 Stages diet and nystatin. When he came back four weeks later, he was taking one quarter teaspoon of nystatin four times a day. Don's moods had leveled out, he had gotten off both his medications, and he had lost three pounds. He had more facial tics but other tics had not returned. He had had some nausea in trying to get off Paxil, but this was better now. He was still obsessive about confessing guilt. Sleep was all right. I saw that his mood was pleasant and he was laughing and smiling. He was blinking his eyes and moving some facial muscles and was moving his head a little. Neurological exam was normal except for the tics.

At this time he was off the medications, on the 4 Stages diet and nystatin, and had actually fewer tics now than while on the medications. I suggested that the treatment be continued.

He came back three months after the first visit. He was still following the diet except for eating a little chocolate on Sundays as a reward (He still had some motor tics and coprolalia.) Emotionally he was doing very well. He had lost about five pounds. He was confessing less often and to fewer things. He was sleeping reasonably well. On exam a few vocal tics were heard. He swung his arms occasionally and he was smiling.

Don came back six months after starting treatment and he said that he was doing fine in school. He was doing well in football. He was keeping up with the other kids much better than last year, and he was getting along with his classmates. The teachers said that things were fine. He was still blinking his eyes and making a few vocal sounds. The force of his vocal tics had decreased, although his tics were worse with stress and varied from day to day. One teacher who had not known him previously had not noticed anything unusual. He was no longer apologizing for everything. On exam a few eye blinks were noted. He was much thinner. No vocal tics were heard.

His parents sent me a Christmas card with the patient smiling in his football uniform.

At the next visit, a year after first being seen, his mother estimated his motor tics were down about 95% and his vocal tics were down 98%. He was no longer confessing to things. He had lost the weight he had gained with the medications. At the appointment, some eye blinking was observed but no other tics. He was on no psychiatric medications and the few tics he had were not interfering with his life.

Three months later, at his most recent visit, his mother reported that the vocal tics were gone and there were only occasional motor tics.

Treating Tourette's takes time. But in the long run, Tourette's responds well to the 4 Stages diet and nystatin. Don is now able to live a normal life, simply by treating his intestinal yeast.

One other Tourette's patient, a star soccer player at the age of 10, had tics so sever that he could not play soccer. He went on nystatin and the 4 Stages diet and was able to play soccer again within a month. He was reluctant to stay on the diet and his mother preferred acidophilus to nystatin, but he was still able to control his tics well enough to play soccer with only partial compliance. Every time he went off the diet significantly, his tics worsened.

Seizures

The 4 Stages diet and nystatin also has been helpful in controlling seizures.

Abby

Abby, 17, told me she had suffered from absence (petit mal) seizures for nine years. She had suffered from migraines for three years. The migraines were now daily. Whatever she ate, she felt abdominal cramping and diarrhea. She was losing weight. She felt exhausted and fatigued. She could be emotionally volatile, laughing one moment, crying the next. Lately she had been having trouble sleeping.

Abby had been treated with anti-seizure medicines for nine years with only minimal effectiveness. The seizures continued. She could not obtain a driver's license. She was in the top 10% of her class but she was slower in doing her homework than she had been earlier in her life. She had to work each night until 10 PM on her homework. Abby reported that the anti-seizure medicine was hurting her school performance. She was concerned about upcoming college placement tests.

In the office, her mood was downcast.

Abby started the 4 Stages diet and nystatin. A month later, Abby reported her stomach problems were gone. Her weight had stabilized and increased. Her migraine headaches were mostly gone. She had a headache now about once per week and when the headaches occurred, they were less severe. Most importantly, the seizures were gone. Her mood was better. Her mind was clearer. She could do her homework all at school now. She was no longer taking anti-seizure medication. In the office, she was smiling and appeared happy. We talked about how she could now obtain a driver's license.

In this case, a child who had suffered for years could now go on to live a normal life, free of seizures, and free of the side effects of medications.

Seizures have many causes, ranging from head trauma to brain tumors to unknown causes. People with known causes for seizures may not respond as well as Abby responded. However, I recommend everyone with uncontrolled seizures of unknown cause try the same treatment as Abby.

Conclusion

All of these patients in all of the cases in this book are real. They all suffered from problems for which their doctors had no answers. They all would have gone to separate specialists for their problems. What other doctor would connect chronic constipation with Tourette's Syndrome, chemical sensitivity with anorexia nervosa?

We can see, though, that all of these patients improved so much that their problems no longer controlled their lives. Don, the boy with Tourette's, could play football and had friends. In only four weeks, Kim no longer had headaches and stomach aches. She no longer was constipated. Deangelo's intestinal gas, which had plagued him for a years, disappeared in three weeks. Norma, who had suffered debilitating headaches, and had difficulty with premenstrual syndrome, was symptom free after seven weeks. Dee could eat again after being anorexic, Tom's chemical sensitivity diminished to the point where he could function. Edna, who could barely get out of bed, had diarrhea and abdominal cramping, was mainly fine in only eight weeks. Nancy, who had been bedridden, was able to get out of bed and function only four weeks after starting anti-yeast treatment. And Abby freed herself from a lifetime of seizures.

Are these cases miracles? Are they aberrations? No. They are cases of real people who were willing to look at their problems in a different way. They were willing to consider the idea, anathema to many people, that what they ate contributed to their health, and to their health problems. When they followed the 4 Stages diet and took nystatin, they healed. They found the extraordinary healing power of anti-yeast therapy. You can too.

Notes

[1] As examples, yeast make acetol, acetoin, 2-methyl-2-pentenal (Suomalainen and Linnahalme, 1966), putrescine and spermine (Stevens, 1981). These chemicals are all listed as skin and eye irritants in the registry of toxic substances, RTECS. Putrescine is extremely destructive to the tissue of the mucous membranes and upper respiratory tract, eyes and skin. It causes burns. Spermine is listed as causing skin and eye burns, and may cause severe and permanent damage to the digestive tract. It causes gastorinestinal tract burns and may cause perforation of the digestive tract.

Suomalainen, H. and T. Linnahalme. Metabolites of alpha Ketomonocarboxylic Acids Formed by Dried Baker's and Brewer's Yeast. *Archives of Biochemistry and Biophysics.* 114:502-513, 1966.

Stevens, L. Regulation of the biosynthesis of putrescine, spermidine and spermine in fungi. *Medical Biology*, 59, 308-313, 1981.

Chapter 4

Autism and Yeast

> *"You know how I had to keep saying the same thought over again and again. This medicine (nystatin) erased all those things and now I don't have to keep saying the same thing over and over."*

This is the statement of a nine year old boy with Asperger's Syndrome (a form of autism) after following the 4 Stages diet and the anti-yeast medicine nystatin.

Children with autistic disorder (defined below) and related autistic spectrum disorders improve when they change their diets to the 4 Stages diet and take the anti-yeast medicine nystatin. In this chapter, I will go through the basics of this treatment and give some possible reasons as to why this treatment works.

First, what is autism or autistic disorder? Autism is a major developmental disorder which affects the ability to communicate and interact with others. The most severely affected children cannot speak. Children who can speak may be limited by an inability to behave appropriately. For example, they may line up toys over and over again to the exclusion of playing with others. Older children may become frustrated, depressed and sometimes aggressive. Autistic persons can injure themselves.

Many children have some of the symptoms of autism. These children receive diagnoses of either autistic spectrum disorder (ASD) or pervasive developmental disorder (PDD).

Parents of children with Autism, PDD or ASD can be deluged with information. They may hear that the cause of autism could be vaccines, the mercury in vaccines, lack of the right vitamin A, immune system deficiencies, the yeast Candida, and casein and gluten opioids (to be explained). Therapies could be chelation, all kinds of supplements, intravenous immunoglobulins, diets free of casein (a protein found in milk) and gluten (a protein found in wheat) anti-yeast diets, antifungal medicines, secretin, as well as standard psychoactive medication to control symptoms. This array of treatment choices is confusing to most parents. How can they sort out what to try, what not to try, and in what order?

My experience as a child psychiatrist who has treated numerous children and adults with autism, and as a parent of a teen with autism, leads me to recommend, as a first line of treatment, changing to the anti-yeast diet and following the treatment I prescribe.

Why? I have treated children and adults who have used standard psychoactive medications, gluten and casein free diets, acidophilus, DMG, secretin, chelation, and other therapies. Simply put, nothing, in my experience, works as well or as quickly to change autistic behaviors as anti-yeast therapy, consisting of the 4 Stages diet and nystatin. Clinically, I know that children all improve when treated with anti-yeast therapy. Furthermore, no parent has ever come back to me and told me that my prescribed therapy did not work or was not helpful. I will explain why below.

Here I will present what I understand about autism and how I treat autism most effectively. I wish to put together the research in such a way that this overwhelming disorder makes at least a little sense to parents and others who may deal with autistic persons. As I said earlier, I have treated dozens of children and adults with autism. When the parents and caregivers follow the treatment plan I outline here, the autism improves. Some children, especially the younger ones, improve to the point that they can function in a regular setting without much support. All improve to some extent.

What causes autism?

Does anyone know for sure what causes autism? The short answer is no. No one knows for sure what causes autism, but more children are developing autistic disorders than in the past. No one knows why this is. However, the fact that nobody has answers about autism's causes does not mean that you cannot treat it. The treatment I prescribe is effective in the long-run. Every family can do it.

The therapy that I use was developed by painstaking observation and treatment of a very severely affected child. The therapy has worked consistently for other children. Based on my observations, I have come to believe that one cause of autism is toxicity or poisoning of the brain. The therapy I prescribe removes from the diet foods containing chemicals which are toxic to the brain and slow it down. The therapy also treats intestinal yeast. This therapy is available to all parents and children, is not expensive, and is non-toxic, but does require a cooperative doctor.

When observing autistic children, I asked the following question: what do these children take in that is toxic to their brains? Toxic in this sense mainly means slowing down the brain so that it does not work well. Perhaps if we could remove these chemicals, these children would do better. In clinical practice, autistic children and adults do better when such chemicals no longer reach their brains. So, we must look to where we can find such chemicals and to how we can remove them.

Yeast chemicals slow down the brain

Before we can understand why preventing such chemicals from reaching the brains of autistic children and adults is effective we need to know a little about some research into observations on brain function in autism. Scientists have studied the brain in autism. Scientists sometimes find an alteration in the size of a brain area. However, other scientists may perform the same study and not find the same result. With one exception, the findings are inconsistent for any dysfunction in autistic brain found so far. There is no current academic unifying explanation for the cause of autism.

Several different scientific groups, however, have made one finding consistently. This finding is that the brain in autism appears slowed down and not as active. In several studies, scientists have found reduced blood flow, especially to the speech areas.[1] Blood flow and activity level of the brain go together.

These findings indicate that the brain in autistic children is less active than in normally developing children. How could this occur? This question is important both for itself and because an answer to this question helps us understand some of the current therapies for autistic children.

One way to look for an answer to this question would be to ask, "What chemicals enter the body and brain of autistic children which would slow down the brain?" Chemicals enter our bodies through the things we eat, breathe in, and come in through our skin. The most obvious source of chemical inges- tion is food. As I described in detail in Chapter 2, and as I will summarize below, food contains chemicals which slow down the brain.

Another possible source of toxic chemicals is from the intestinal yeast, Candida albicans. Candida is present in greater quantities after people take antibiotics.[2] This yeast makes a whole variety of toxic chemicals. These chemicals include toxic alcohols,[3] acetone[4] (which causes coma[5]), ethyl acetate[6] and the nerve poison hydrogen sulfide.[7] All of these chemicals slow down the brain. [8] In addition, ethanol (the alcohol in alcoholic beverages) is known to disrupt brain development at all phases of development.[9]

More chemicals come from inside the gut. Chemicals made by Clostridia, a bacteria found in the gut after antibiotic use,[10] makes chemicals which slow down the brain.[11]

These chemicals all are being made inside the intestinal tract of any child who has excess yeast. These chemicals are abundant in any person who has taken antibiotics. I compare all these chemicals to having a leaky toxic chemical plant inside from which the child cannot escape. The toxic chemicals are right inside of the child.

In other words, yeast chemicals, chemicals from food, and from intestinal Clostridia could account for some of the brain slowdown in the studies I discussed above.

Let us look at food. Many toxic chemicals come in from food. For example, one of the most common food additives is malt, a cheap sweetener. One can find malt in almost all packaged and processed foods, including commercial white flour. Malt contains chemicals called pyrazines [12] which will slow down the brain.[13]

Another common food, found in most sauces, condiments and salad dressings is vinegar. Vinegar contains ethyl acetate. In Chapter 2, I discussed many such foods and chemicals in great detail. These chemicals slow down the brain.

So why should we take these foods away? Many people argue that children who have autism deserve to eat whatever they want, because life is so difficult for them. My answer is, one reason life is so difficult for them is because they eat such foods. If we remove the foods that cause problems, these children will have an easier time. I truly believe that if autistic children could talk to us, they would say, "Do not slow down my brain. I have a hard enough time functioning already."

In other words, many chemicals could be affecting the brains of autistic children. These chemicals slow down the brain and disrupt development. Think of these chemicals collectively as being an internal leaky toxic chemical plant.

We can remove these chemicals and see what happens to brain functioning. As a doctor, I have seen time after time that when we treat for yeast by removing toxic chemicals coming into the body and by killing the yeast within the body, autistic children and adults improve significantly.

Before presenting cases demonstrating this improvement, I will explain how understanding these chemicals' effects on the brain helps us understand some difficult behaviors seen in autistic children.

What is the effect of slowing down the brain?

What is the effect of slowing down the brain? These chemicals do not just slow down the brain, they also slow down the nerves. They throw off sensation. If you think of the nerves as carrying basically two types of information, pleasure and pain, you can understand this better. The signals carrying information about pain are much harder to dampen than signals carrying information about pleasure. For example, if you undergo surgery, you need higher doses or more powerful anesthetics to block pain than to block pleasure.

We can see this same effect in children with autism. They are taking in chemicals that numb the nerves. The easiest sensations to numb are pleasure. After the pleasure sensations are numb, the pain sensation is intact. Thus a light touch, which may feel good to a person with intact nerves, feels painful to a child with autism in whom the majority of the nerves are slowed down and numbed up. This often is referred to as tactile sensitivity, and whole therapies are devoted to decreasing tactile sensitivity.

This altering of sensation may also affect the mouth, which is why I believe these children become picky eaters. New foods may not feel right to them. In my experience, food choice improves with the anti-yeast diet. I have had some very severe cases where children had virtually stopped eating. The parents began the 4 Stages diet and their children were able to eat again.

These chemicals which numb the nerves may also lead to such disordered sensation that feelings of pain may feel better than nothing at all. For example, think of how you feel after receiving Novocain for a dental procedure. You can't feel part of your face. You feel strange. Pain may be preferable to this type of feeling for autistic children, leading to self injurious behavior.

Besides numbing up the nerves, autistic children and adults can suffer from all the same kinds of problems and pains other people with Candida suffer from. For example, many autistic children have skin rashes, constipation and/or diarrhea. Autistic children and adults can suffer from headaches and abdominal pain. Sometimes this pain can be chronic and severe. When autistic children or adults scream, there is a reason. They can be in severe pain.

I will give you some cases:

Cases of autistic children treated with the anti-yeast diet and nystatin

Anton

Anton, 2 years and 2 months, was seen in my office one month after receiving a diagnosis of autistic spectrum disorder. His parents reported the following history: At 18 months he had stopped talking. He had received many antibiotics from birth on for a kidney condition. He had also had many ear infections. He had ear tubes placed six months previously, which had stopped the ear infections. Antibiotics had been stopped for the kidney problem about the same time. He was now enjoying watching things spin in circles. He liked to line things up. He was having tantrums for what he wanted. He did not like to communicate with his parents, and he did not respond to his name or to his mother. He had poor eye contact. He had been banging his head. Anton was not sleeping well. He was waking during the night. He had rashes on his face and chest and up to four loose stools per day. He had had thrush (yeast in his mouth) when younger. The family had not tried any therapies except applied behavioral analysis therapy.

When I saw Anton in my office the first time, he did not respond to his name. He counted, both spontaneously and with prompting. He was calm. I put Anton on the 4 Stages diet and nystatin.

Two months later, Anton was sleeping through the night. He was more alert and his skin had cleared up. During the treatment, Anton's parents had taken him off milk and wheat (Stage III of the 4 Stages diet). Anton's diarrhea had cleared up. One of Anton's teachers had noticed a change. He was babbling more and making more eye contact. He could count to twenty and say ABC's. He was repeating some words, but he still had a hard time getting words out. He was interacting more. He was coming to his parents for hugs and kisses. He was responding to his name some of the time. At my office, he babbled and played appropriately. At home, he still had tantrums.

After five months on the 4 Stages diet and nystatin, Anton's difficult behaviors had been reduced. He no longer banged his head at the start of behavioral therapy. He was speaking more when playing with other children in structured settings. He could say some words when playing, such as "run" and "jump". He was singing little songs. He was saying a fair number of words and was responding to his name 70% of the time. He liked to count. He could recognize letters (but not words). He had maintained the improvement with bowel function. In addition to the diet and nystatin, I also started Anton on a homeopathic remedy called Natrum muriaticum and prescribed topical secretin.

At this same visit, Anton's parents reported that he was progressing steadily. He could say please, thank you, excuse me. He was not echoing. He was animated and interacting more. Father said that he looked forward to coming home to see how Anton is developing new skills. At this visit, Anton responded to his name. He said words such as "bye" and "go outside" and "all done". He smiled and made eye contact. A speech evaluation two months prior to this appointment noted that Anton was now saying words such as "hug, kiss, bubble, up/down, ball, dad, bye bye, open the door, more, choo choo, all done, there, march, boom." The speech therapist reported, "Emerging use of words to describe current activity."

Anton's case shows how a nonverbal child developed beginning verbal skills and dramatically improved his behavior and functioning after only a few months of treatment.

David

David came in at the age of five with a diagnosis of autism given at age two and a half. At 18 months, he had lost his language. He could say the alphabet, but no words. He went to a number of doctors and received a diagnosis of autism about a year later. David was having subclinical seizures on EEG (electroencephalogram, a test of brain electrical activity). DMG (dimethyl glycine) calmed him and helped his mood. Some other supplements had been helpful. He had started applied behavioral therapy at age three. He learned some reading through the association method.

David had had an elevated lead level at 18 months at a prior residence. Mother tried homeopathic remedies but could not find anyone to do chelation therapy (a procedure to take toxic minerals out of the body). The family moved and the lead level went down in six months. He had been continued on supplements and behavioral therapy. He had had loose stools as an infant. He had never had an ear infection. He was taking Depakote, an anti-seizure drug, at 125 mg. twice a day. He had never had an obvious seizure.

David walked on time. Some speech had begun with behavioral therapy.

When I saw David, he had a rash on his buttocks. In the room he was active and responded to some questions with rote answers. He had echolalia, a sign of autism which means the repeating of phrases said to a child. He was not destructive.

I prescribed the 4 Stages diet and nystatin.

A month later, following only Stage I of the diet, David was talking more spontaneously without prompts. He could nod yes or no. He was relying less on body language. His behavior was more spontaneous. He was described as acting naughtier, more like a "normal kid." He was eating better and was in a growth spurt. He was trying new foods with less difficulty. He was not running away from the dinner table. He was helping to set the table. His appropriate play had increased. He was interacting with his brother and was generally more interactive and could say, "that's mine." He was taking a quarter teaspoon of nystatin four times a day as well as Depakote.

In my office, David played appropriately and he showed me his toys. He was smiling spontaneously and appeared comfortable. He fought with his brother over toys and appeared more normal. The rash on his buttocks was better.

David came back six months later. His mother reported that he was doing much better. He had been off Depakote for two months. He cried more easily, and he hit if he was frustrated. This was more normal behavior for his age. David seemed more clear headed and more "present" (not withdrawn). He was very verbal, saying appropriate things. David was responding to what other people were saying. He would greet people spontaneously. David was making appropriate comments to others. He was working on these things in therapy. He said phrases to himself.

Mother said that David looked more "normal." He was more easily frustrated now.   He had filled out. He no longer had dark circles under his eyes. He could read and write. On exam he made appropriate comments and repeated some phrases. Overall he was much more verbal. He responded to questions and played appropriately. His skin was clear. He liked to take the nystatin. He was refusing dairy.

A year after beginning treatment, David was in a full day kindergarten shadowed by a therapist. He was eating better, was growing well, was not pale, and was more spontaneous. He no longer echoed, although he still reversed pronouns. He had trouble with ordinal counting, such as first and last and with abstract concepts. He played very appropriately with toys. He skin continued to be clear. He was taking one quarter teaspoon of nystatin four times a day and was on Stage 1 of the diet plus no dairy. In my office, he answered questions appropriately, played appropriately, smiled and interacted with his mother. His mother also reported that David could verbalize when he had pains, which he could not do before. He could toilet himself and dress himself, and he was not running away. I commented to mother, "One had to look closely at him to see that there was anything wrong."

I next saw David several years later at the age of 8. He had stayed partly with the treatment program. He now no longer was diagnosable as autistic. He had some speech and language problems and some social delays. The other kids had no idea how impaired he had been and they were treating him like any other child.

This case shows that even following the first stage of the 4 Stages diet, combined with nystatin, can bring a young child out of autism. David had other therapy, to be sure. In my experience, the behavioral and educational therapies are much more effective if the child's internal systems are functioning well.

Daniel

Daniel came in at the age of 3 with a diagnosis of high functioning autism. He was delayed in speech reception and processing. He had been in speech and occupational therapy for a year. He was being treated for sensory integration issues, and he was limited by his speech. Daniel was normal at birth, but had had many ear infections, even after ear tubes at 18 months. He was not sleeping through the night. He frequently woke at night, and one of his parents had to stay with him for him to fall back asleep. At 18 months, he was speaking and putting two words together. However, he did not go on to making sentences and he used the same phrases over and over. He did not ask for things, but instead would pull his parents to what he wanted.

Hearing was fine. A speech evaluation had shown him 11 months behind in speech. An occupational therapy evaluation showed him at the 19 month old level. (He must have been around two and half at this time). Now he was talking more and answering some questions. His mother had put Daniel on Stage I of the 4 Stages Diet five days prior to the appointment. Since changing his diet, he had been having good days without battles. His tantrums were diminished.

On exam he did not respond to requests or questions. He answered questions when his mother asked about his toys. He produced a few short sentences about his play. He made little eye contact. He was under the chair at first. Later, he became agitated as he was trying to climb on the shelves in the room.

Daniel started on nystatin and continued on the 4 Stages diet.

Daniel came back six weeks later. He was now on Stage 2 of the diet and was going off milk. He had clearly improved. He was sleeping through the night. His tantrums were minimal. He was making great eye contact and was much more interactive. He asked appropriate questions and gave appropriate answers. His appetite had improved. He was interacting with other children at school. His mother said that they were "on a roll". He was taking one fourth teaspoon of nystatin four times a day. In my office, he interacted and answered questions. He pointed to his mother when requested and he played appropriately and smiled while playing. His mood was pleasant.

He still had eczema (not mentioned at first visit) and complained of belly pain. He was still eating wheat.

Several weeks later Daniel's urine was tested for casein and gluten peptides. The test came back positive, and his mother took him off casein and gluten. (I discuss the significance of casein and gluten free diets in a few pages.)

Daniel came back four months after the first visit and he was doing well. Daniel had continued to improve, but he was anxious and shy. He was not nervous if he was in a group, but was shy if singled out. He was trying to be interactive. He wanted to play with the other kids but seemed not to know how. He was much more interested in the world around him. He did not want to be alone and asked to play with the boys. He could answer questions appropriately and his sentences were more complicated. He was doing well in speech therapy and he needed less sensory input. He was swinging and jumping less. His sleep was fine. According to his mother, he seemed to be aware that he was feeling better. He was taking half a teaspoon of nystatin powder twice a day. I saw that his rash was clearing. In my office, Daniel played quietly. He seemed shy but made eye contact and softly answered questions. He was talking as he played.

Overall, in less than a year, Daniel had gone from not sleeping, not being able to speak well or put sentences together, not responding to request and questions, to playing more appropriately, talking more appropriately, sleeping, answering questions and making eye contact.

These cases are not sensational or atypical. Anton, David and Daniel show the incredible improvement autistic children can make when following the 4 Stages diet and taking nystatin.

How does a gluten and casein free diet fit into the anti-yeast diet?

Parents who research autism on the internet often will turn up the "gluten and casein free diet." Gluten is a protein found in wheat, barley, oats, and rye, and casein is a protein found in cow's milk and dairy products. How can we understand this treatment? Where does this therapy come from? How does it fit into anti-yeast treatments?

Let me start by answering the third question first. I prescribe removing gluten and casein from the diet in Stage III of the 4 Stages diet. I consider removing casein and gluten part of anti-yeast therapy, as I will explain later.

But for now let us go back to the first two questions: How can we understand the gluten and casein free diet, and where does this therapy come from? To answer these questions, we must look at the brain's endorphin system. These pathways affect the brain's activity. First let me explain the endorphin system.

Endorphins slow down the brain

The endorphin system in the brain is complicated. Endorphins are produced internally. They are part of a large class of substances called "opioids," so called because they activate the same structures as the opiate drug morphine.

Scientists know that when morphine is injected into an experimental animal, morphine binds to brain opioid receptors, and this binding leads to decreased metabolic rate.[14] The metabolic rate, in simplistic terms, is how active the brain is. This decreased metabolic rate will lead to decreased blood flow. People also have observed that autistic children show decreased brain activity and have searched for answers to explain this finding.

One leading answer involved understanding endorphins. Endorphins react at the brain's own internal opioid receptors.

Such receptors are found in many brain areas, including temporal lobe and limbic structures, important structures for speech and regulation of emotions.[15]

The casein/gluten free diet is based on the observation that gluten and casein contain opioids, which activate endorphins and slow down the brain. However, one should understand that other reasons exist for increased endorphin activity in the brain in addition to gluten and casein peptides found in the diet.

Yeast chemicals will release endorphins

When the brain is faced with any kind of toxic chemicals coming in, one of the responses is to release endorphins. How do we know this? Endorphins, the body's internal opioids, are released under stress.[16] They are also released when anesthetic agents are given.[17] Ethanol, the alcohol in alcoholic beverages as well as a major yeast chemical, causes endorphin release.[18]

The chemicals which yeast produces, discussed in Chapter 2, include toxic alcohols, acetone and hydrogen sulfide.[19] We know that these chemicals, similar to anesthetic agents, slow down the brain and nerves.[20] Most likely, other yeast alcohols as well as acetone and hydrogen sulfide, will cause endorphin release. In other words, many yeast chemicals cause endorphin release, which will then lead to decreased brain activity.

Dietary opioids from milk and wheat will slow down the brain

The other source of opioids is external, in the diet. These opioids are what the gluten/casein free diet is all about. When the body's digestive system degrades casein from milk proteins and gluten from wheat and other grains, structures remain which have opioid activity.[21] These structures are difficult for the body to degrade.[22] These compounds are absorbed and can cross the blood brain barrier.[23] No one knows how the brain clears these compounds. Opioids have been found in the urine and cerebrospinal fluid of schizophrenic adults .[24] Similar peptide profiles are seen in urine of autistic children.[25] Presumably such peptides activate endorphin sites in the brain. Such peptides could be one factor accounting for decreased brain activity seen in the brains of autistic children.[26] In fact, removing casein and gluten from the diets of autistic children usually leads to improvement.[27]

Overall, many chemicals will slow down the brain.

In other words, many sources of chemicals slow down the brain. There are yeast chemicals and there are dietary chemicals. There are opioids from the diet and internal opioids released from stimulation by alcohols coming in, such as yeast alcohols. These chemicals affect the brains of autistic children. We don't know how much of the decreased brain activity is due to yeast chemicals and how much is due to casein and gluten.

The exact number really is not important when thinking about how to treat people. These chemicals all have some effect and should be considered when setting out your child's treatment plan. The only question, as I will discuss below, is in which order to attack the problem, given the overwhelming nature of autism and the confusing nature of autism treatments. Reducing the intake of all of these chemicals leads to improvement in autistic persons.

Autism is a complicated disorder and other processes must be going on which contribute to its development. However, understanding yeast chemicals and endorphins is very important. Why?

First, because historically, autism appears related to introducing antibiotics into medical practice. Autism was first described by Dr. Leo Kanner in the United States in 1943,[28] and by Dr. H. Asperger in Austria in 1944. [29] The first major antibiotic was introduced into medical practice in the mid to late 1930's.[30] I do not believe this is coincidental. I have spent considerable time as well reviewing older medical textbooks from the 19th century. In those books, there is no mention of the kinds of behavioral and developmental problems that we see starting in the early 1940s, and described by Dr. Kanner and Dr. Asperger.

The second reason for taking yeast seriously is that you can do something about the problem by reducing intake of both yeast chemicals and opioid chemicals. When children reduce their intake of these chemicals, they improve significantly. Their brains function and develop better. This is an easy, inexpensive treatment that works.

Setting treatment priorities

We know that people talk about two "diets" recommended for autistic persons in addition to other types of treatment. One diet is the anti-yeast diet, and the other is the gluten/casein free diet. Are these two diets mutually exclusive? No. I consider them part of the same treatment, not two separate diets. In my mind, the only question is which part to start first and which to add later.

When working with patients I first recommend reducing yeast chemicals, then reducing gluten and casein. This recommendation is based on my clinical experience. Autistic children do best if yeast chemicals are removed as soon as possible. One usually can see changes in 2-6 weeks. Children who are under two usually can be turned around completely, and led to a path of normal development. After the age of four, improvement is slower, but still can markedly change the quality of life of a person with autism. Even adults with autism show major behavioral improvement when started on the 4 Stages diet. Like the autistic children, the adults are in less pain and can handle things better after they follow the 4 Stages diet.

In my clinical experience, children who start on the gluten/casein free diet usually see much slower, inconsistent results than children who first start the yeast free diet. There are exceptions, of course. These slower changes can be discouraging to the parents who are investing so much time and effort into changing their child's diet, and parents may give up before they can see real results.

Another reason for eliminating gluten and casein after eliminating foods containing toxic yeast chemicals is that many highly toxic foods are permissible on a gluten/casein free diet. Toxic foods that are permissible on the standard gluten/casein free diet, but are known to cause problems, include vinegar, pickles, malt, peanut butter, chocolate and soy sauce. For example, the main milk substitute recommended on the gluten/casein free diet has maltodextrin as a main ingredient. A large industry is devoted to making processed gluten/casein free foods. From reading the labels of such foods, I found that many of the gluten/casein free foods contain ingredients known to cause problems in autistic persons, including malt, vinegar, peanuts, and chocolate. Usually, gluten/casein free diets recommend eliminating barley malt because barley contains gluten. However, they permit malt from "non-gluten" sources.

This is erroneous. All malt starts with barley. Food processors then mix malt with other foods. The result is maltodextrin from corn, tapioca, potatoes, or other foods. One of the most common questions I get is whether these other sources of maltodextrin are OK. The answer is no. So, beware of labels which say "maltodextrin (corn)". Maltodextrin is malt mixed with corn syrup, potato starch, or some other ingredient. The name of the other ingredient, for example, corn, is placed after the "maltodextrin" on labels to show that the maltodextrin contains corn. The maltodextrin also contains malt which comes from barley.

So eliminating gluten and casein only removes a small category of problematic foods, which may be why the casein/gluten free diet works slowly and inconsistently in most people. Removing yeast and fermented foods recommended in Stages I and II of the 4 Stages diet leads to more consistent, faster changes in the patient's functioning. By clearing out the worst toxic foods first, the patient lays the foundation for eliminating gluten and casein, thus improving much more quickly.

In my clinical experience, when gluten and casein are removed from a patient's diet after yeast chemicals are removed, I usually see more dramatic results than when it is removed before the yeast chemicals. Leaving toxic foods in the child's diet usually slows the child's progress or leads to inconsistent results. One aspect of the child's development improves while others get worse. For example, a typical result is that speech improves slightly, but sleep gets worse and tantrumming increases.

When setting treatment priorities, I recommend starting with Stages I and II of the 4 Stages diet to eliminate the most toxic foods. Then I recommend going on to Stage III, to eliminate casein and gluten. Finally, I recommend, if appropriate in a given case, moving on to Stage IV to eliminate the last group of problematic foods.

Use of the opioid blocker naltrexone

I previously discussed how activity in the brains of autistic children seems to be reduced. This finding has led to the therapies of yeast free and gluten/casein free diets. We can see that if many toxic chemicals come in, too much release of internal endorphins may also be occurring. This may slow down the brain.

This observation has also led to using a medication, naltrexone, to block opioids in the brain. Giving naltrexone may allow the brain to be more active.

The studies on using naltrexone in autistic persons have shown inconsistent results. These studies all have used high doses, and minimal results are seen. I have found that low doses work better, and then only with the diets recommended above. I do not recommend giving an opioid blocker if the patient is still eating numerous dietary opioids. Properly used with the correct doses, naltrexone can be effective. I have seen positive results with doses in the range of one to six milligrams per day.

Other treatment issues

To date, I have never had a parent of an autistic child come back to me and say that they had done exactly what I prescribed and their child had not improved. All autistic children I have seen who have followed the anti-yeast diet and taken nystatin have improved. With this treatment you see consistent gradual improvement. The treatment usually eliminates the need for "medications" to control behavior. Skin problems, constipation, diarrhea and other medical problems usually improve or disappear.

However, several things can interfere with treatment. First, food sensitivities can interfere. If a food to which the child is sensitive is substituted for the excluded foods, the child can have acute reactions to this new food. The most common offender here is corn, which is not excluded on casein free/gluten free and it is not excluded in Stage I of the 4 Stages diet. Corn is gluten free and often is increased or substituted for wheat; however, corn can be very mold contaminated. Milk can also be such an offender.

Other treatments can also interfere. One of these treatments is high dose vitamins. I have received many inquiries saying that the child is taking high doses of vitamins and is on the gluten free/casein free diet. Then nystatin is added and does not help. I have seen a number of such children who also are taking both high doses of vitamins and nystatin. In some of these children, nystatin seems not to work.

If this happens to your child, I would suggest stopping all supplements for a few weeks, then adding them back one at a time to determine if the vitamins or supplements are interfering with anti-yeast treatment, or whether such supplements are even helpful or necessary. Parents who have done this often have found that the supplements were not helping. Currently, I recommend that no vitamins or supplements be used when starting anti-yeast therapy. There is no reason to use high doses of vitamins together with nystatin and anti-yeast therapy.

Why might all these supplements not be helpful? The nutritional supplements are concentrated sources of nutrients to assist in cell growth. What gets first crack at them? The yeast gets them first. The supplements may help the yeast grow and make it much harder to clear out the yeast. Yeast is very tenacious. Use of vitamins and other supplements make it that much harder, if not impossible to clear out yeast. I recommend that no supplements be used when treating for yeast.

Another problem patients encounter happens when they take nystatin without changing the diet. Nystatin only works with the anti-yeast diet. Taking nystatin can have bad results without changing diet. Without changing diet, the yeast keeps coming back. In these cases, the yeast level may be going up and down, which is quite hard for the patient. I discuss other issues more thoroughly in Chapters 16 and 17, including use of acidophilus.

Pregnancy after birth of an autistic child

Couples who have an autistic child are at a one in fifty risk of having another autistic child. This is of course discouraging. We don't know why this is true. This may be purely genetic or due to environmental influences, or a combination of both. One factor may be previous antibiotic use by the mother. These women may be carrying yeast in their intestinal tracts. The yeast make toxic chemicals which may affect the brain of a fetus in utero. I emphasize that we don't really know for sure what the reason is for the increased risk of autism. However, in this case, one might be able to prevent possible problems by treating the mother for yeast before pregnancy as a precautionary measure. This may reduce the risk of autism in a later pregnancy. After the child is born, I recommend avoiding antibiotics to the greatest extent possible. This means that if mother is nursing the child, she should avoid antibiotics if possible. If the child needs antibiotics, give nystatin at the same time to avoid problems. There are no guarantees, of course, but this is one possible way to reduce risk without incurring risk in the mother. As with all medical issues, be sure to consult your personal health care provider.

Notes

[1]George, M. S., Costa, D. C., Kouris, K., Ring, H. A., and P. J. Ell. Cerebral blood flow abnormalities in adults with infantile autism. *Journal of Nervous and Mental Disorders.* 180:413-417, 1992; Goldberg, M.J., Mena, I., Miller, B. And C. Thomas. Neurospect Findings in children with "Autistic Syndrome". *Proceedings of the 1996 Autism Society of America National Conference.* 226-228, 1996; Mountz, J. M., Tolber, L.C., Lill, D.W., Katholi, C.R., Lin, H.G. Functional deficits in autistic disorder: characterization by technetium-99mHMPAO and SPECT. *Journal of Nuclear Medicine.* 36(7):1156-62, 1995.

[2]Giuliano, M., Barza, M., Jacobus, N.V. and S.L.Gorbach. Effect of broad spectrum parenteral antibiotics on composition of intestinal microflora of humans. *Antimicrobial Agents and Chemotherapy.* 31:202-206, 1987.

[3]Berry, DR., and D.C. Watson. Production of organoleptic compounds. In *Yeast Biotechnology*, Berry, DR., Russell, I and Stewart, GG., Eds., London, Allen and Unwin, 1987, p. 345; Lingappa, B.T., Prasad, M., Lingappa, Y., Hunt, D.F., and K. Biemann. Phenethyl Alcohol and Tryptophol: Autoantibiotics Produced by Fungus Candida albicans. Science. 163:192-193, 1969; Rankine, B.C. Formation of higher alcohols by wine yeasts. *Journal of the Science of Food and Agriculture.* 18:583-589, 1967.

[4]Berry, DR., and D.C. Watson. Production of organoleptic compounds, in *Yeast Biotechnology*, Berry, DR., Russell, I and Stewart, GG., Eds., London, Allen and Unwin, 1987, p. 345; Lingappa, B.T., Prasad, M., Lingappa, Y., Hunt, D.F., and K. Biemann. Phenethyl Alcohol and Tryptophol: Autoantibiotics Produced by Fungus Candida albicans. *Science.* 163:192-193,

1969; Rankine, B.C. Formation of higher alcohols by wine yeasts. *Journal of the Science of Food and Agriculture.* 18:583-589, 1967.

5Acetone causes gastrointestinal irritation, narcosis, injury of the kidney and liver and can cause coma. From Toxicology of Drugs and Chemicals by W. B. Deichmann and H. W. Gerarde, Academic Press, New York, 1969, p. 64. Acetone causes nausea and vomiting, gastric hemorrhage, CNS sedation, respiratory depression, ataxia, paresthesia and ultimately coma from *Principles of Clinical Toxicology*, 2nd edition by T. A. Gossell and J. D. Bricker, Raven Press, New York, 1990, p.80.

6Torner, M. J. Martinez-Anaya, M. A., Antuna, B., Benedito de Barber, C. Headspace flavour compounds produced by yeasts and lactobacilli during fermentation of preferments and bread doughs. *Int. J. Food Microbiol.* 15(1-2):145-52, 1992.

7Rankine, B.C. Hydrogen Sulphide production by yeasts. *Journal of the Science of Food and Agriculture.* 15:872-877, 1964; Berry, DR., and D.C. Watson. Production of organoleptic compounds. In *Yeast Biotechnology*, Berry, DR., Russell, I and Stewart, GG., Eds., London, Allen and Unwin, 1987, p. 345.

8Gillman, AG., Goodman, L.S. and A. Gilman. *Goodman and Gilman's The Pharmacological Basis of Therapeutics.* 6th ed. Macmillan Publishing Co. New York, 1980, c. 18; Gosselin, RE., Hodge, H.C., Smith R.P. and MN Gleason. *Clinical Toxicology of Commercial Products.* 4th edition, sect. III, PP.169-173 and 271-274.

9 For example, ethanol exposure during the third trimester equivalent in rats depletes major brain cells (West, JR and Goodlett, C.R.. Teratogenic Effects of Alcohol on Brain Development. *Annals of Medicine.* 22:319-25, 1990.). If the liver is not mature enough to metabolize these alcohols, development and function of brain could be affected. Unfortunately, this process could start in utero. No information is available on the liver's capacity to metabolize these alcohols and H_2S in infants.

10George, WL., Rolfe, RD. and Finegold, S.M.. Clostridium difficile and its cytotoxin in feces of patients with anti-microbial agent associated diarrhea and miscellaneous condition. *J. Clin. Microbiol.* 15:1049, 1982.

[11]Clostridium synthesizes p-cresol and other Clostridium species synthesize phenol, (Elsden, et al., 1976). Phenol is used as an antiseptic and kills bacteria. Both phenol and p-cresol are toxic CNS depressants (Gosselin, et al, 1976, and Deichmann and Keplinger, 1981). The effects of phenol and p-cresol would be additive with the yeast alcohols and H_2S. Elsden, S., Hilton, MG and JM Waller. The end products of the metabolism of aromatic amino acids by Clostridia. *Archives of Microbiology.* 107:283-8, 197; Deichmann, W. B., and Keplinger, ML. Phenols and phenolic compounds in *Patty's Industrial Hygiene and Toxicology*, V.2A, John Wiley and Sons, New York, 1981, c.36, pp.2597-2601; Gosselin, RE., Hodge, H.C., Smith R.P. and MN Gleason. *Clinical Toxicology of Commercial Products.* 4th edition, sect. III, PP.169-173 and 271-274.

[12]Hough, J. S., Briggs, D. E., Stevens, R. and T. W. Young. Brewery Fermentations. *Malting and Brewing Science*, 2nd ed., Chapman and Hall, London, vol. 2, pp. 462-471, 1982.

[13] Nishie, K., Waiss, A. C. Jr., and A. C. Keyl. Pharmacology of Alkyl and Hydroxyalkylpyrazines. *Toxicology and Applied Pharmacology.* 17:244-249, 1970.

[14]London, ED, Broussolle, E.P., Links JM, Wong, D.F., Cascella, N.G., Dannals, R.F., Sano, M., Herning, R., Snyder, F.R., Rippetoe, L.R.. Morphine-induced metabolic changes in human brain. Studies with positron emission tomography and [fluorine 18] fluorodeoxyglucose. *Archives of General Psychiatry.* 47(1):73-81, 1990.

[15]Mansour, A, Wason, S.J., and Akil, H. Opioid receptors: past, present and future. *Trends in Neurosciences.* 18(2):69-70, 1995

[16]Khachaturian, H., Lewis, ME, Schafer, M.K.., and Watson, S.J.. Anatomy of the CNS opioid systems. *Trends in neurosciences.* 8:111-119, 1985.

[17]Cork, R.C., Hameroff, S.R., and J.L. Weiss. Effects of halothane and fentanyl anesthesia on plasma beta-endorphin immunoreactivity during cardiac surgery. *Anesthesia and Analgesia.* 64(7):677-80, 1985; Maiewski, S., Muldoon, S., and Mueller, G.P. Anesthesia and stimulation of pituitary beta-endorphin release in rats. *Proceedings of the Society for Experimental Biology and Medicine.* 176(3):268-27, 1984.

18Seizinger, BR., Bovermann, K, Hollt, V. And A. Herz. Enhanced activity of the Beta endorphinergic system in the anterior and neurointermediate lobe of the rat pituitary after chronic treatment with ethanol liquid diet. *The Journal of Pharmacology and experimental Therapeutics.* 230:455-461, 1984.

19See footnotes 3 through 7.

20Gillman, AG., Goodman, L.S. and A. Gilman. *Goodman and Gilman's The Pharmacological Basis of Therapeutics.* 6th ed. Macmillan Publishing Co. New York, 1980, c. 18; Gosselin, RE., Hodge, H.C., Smith R.P. and MN Gleason. *Clinical Toxicology of Commercial Products.* 4th edition, sect. III, PP.169-173 and 271-274.

21Zioudrou, C., Streaty, R. A., Klee, W. A. Opioid peptides derived from food proteins. *Journal of Biological Chemistry.* 254: 2446-2449, 1979.

22Gardner, M.S. Intestinal Assimilation of intact peptides and proteins from the diet-a neglected field. *Biological Reviews.* 59:289-331, 1984.

23Hemmings, WA. The entry into the brain of large molecules derived from dietary protein. *Proceedings of the Royal Society of London - Series B: Biological Sciences.* 200(1139):175-92, 1978; Hemmings, C. Hemmings, WA., Patey, AL., and C. Wood. The ingestion of dietary protein as large molecular mass degradation products in adult rats. *Proceedings of the Royal Society of London - Series B: Biological Sciences.* 198(1133):439-53, 1977; Ermich, A., Ruhle, H-J., Neubert, K. et al. On the blood-brain-barrier to peptides: 3H beta-casomorphin-5 uptake by eighteen brain regions in vivo. *Journal of Neurochemistry.* 41:1229-1233, 1983.

24Lindstrom, L.H.., Besev, G., Gunne L.M., and Terenius, L. CSF Levels of Receptor-Active Endorphins in Schizophrenic Patients: Correlations with Symptomatology and Monoamine Metabolites. *Psychiatry Research.* 19:93-100, 1986; Hole, K., Gergslien, H., Jorgensen, HA. And O.G. Berge, Reichelt, KL, and O.E. Trygstad. A peptide-containing fraction in the urine of schizophrenic patients which stimulates opiate receptors and inhibits dopamine uptake. *Neuroscience.* 4:1883-93, 1979.

[25] Reichelt, K L., Knivsberg, A M., Lind, G., and M. Nodland. Probable etiology and possible treatment of childhood autism. *Brain Dysfunction.* 4:308-319, 1991.

[26]George, M. S., Costa, D. C., Kouris, K., Ring, H. A., and P. J. Ell. Cerebral blood flow abnormalities in adults with infantile autism. *Journal of Nervous and Mental Disorders.* 180:413-417, 1992; Goldberg, M.J., Mena, I., Miller, B. And C. Thomas. Neurospect Findings in children with "Autistic Syndrome". *Proceedings of the 1996 Autism Society of America National Conference.* 226-228, 1996; Mountz, J. M., Tolber, L.C., Lill, D.W., Katholi, C.R., Lin, H.G. Functional deficits in autistic disorder: characterization by technetium-99mHMPAO and SPECT. *Journal of Nuclear Medicine.* 36(7):1156-62, 1995.

[27] Knivsberg, A. M., Reichelt, K. L., Nodland, M., and T. Hoien. Autistic syndrome and diet: a follow-up study. *Scandinavian Journal of Educational Research.* 39(3):223-236, 1995.

[28] Kanner, L., "Autistic Dusturbances of Affective Contact." *Nervour Child* 2:217-250, 1943.

[29] Aspberger, H. "Die autistichen psychopathen im Kindesalter." *Archiv fur Psychiatrie und Nervenkrankheiten*, 17:76-136, 1944.

[30] Mandell, G.L. and Sande, M.A. Antimicrobial Agents: Sulfonamides, Trimethoprij-Sulfamethoxazole, and Urinary Tract Antiseptics, in Goodman and gilman, *The Pharmacological Basis of Therapeutics, 6th ed.* (1980), 1106.

Chapter 5

ADHD, ADD and Yeast

Children with ADHD (attention deficit hyperactivity disorder) and ADD (attention deficit disorder) concentrate much better after they are treated for the intestinal yeast Candida albicans. We will see why this might be in a moment. First, what is ADHD/ADD and what do we know about its cause?

No one knows the cause of attention deficit hyperactivity disorder, abbreviated ADHD, or of attention deficit disorder, abbreviated ADD. The difference between ADHD and ADD is the "H" for hyperactivity found in children with ADHD. For convenience in this chapter, I refer to both ADD and ADHD as "ADHD."

Both disorders share common characteristics: difficulties with concentration, staying on task, focus, and distractibility. Children cannot sustain mental effort. The difference between the two disorders is the "H" for "hyperactivity." By definition, children with ADHD are hyperactive, meaning they cannot sit still in settings in which they would be expected to sit still, such as in school.

The brain in all of these children seems almost tired or asleep and unable to sustain effort for more than a short time.

One major study of the brain supports the idea that in ADHD, the brain is slowed down. This study from the National Institute of Mental Health is a landmark study on ADHD. This was a study of adults with ADHD who never had taken medicine for the disorder.

The researchers gave these adults a radioactive compound which resembles glucose, the sugar the brain uses for energy.[1] The researchers also gave the same radioactive compound to adults that did not have ADHD. They measured the uptake of this radioactive compound. The more the compound is taken up by the brain, the more active the brain is.

The researchers found that many areas of the brains of the ADHD group took up less of the energy compound than the brains of the non-ADHD group. These brain areas, thirty out of the sixty measured, were less active in the brains of people with ADHD. The researchers more significantly found a tendency of lower brain activity in the brain areas which control higher thought processes.

Based on these results, we can think of the brain in ADHD as slowed down and not as active. The academics have no answer for why they find such decreased activity in the brain.

In a later study, the same researchers showed that giving the drug Ritalin, commonly prescribed for ADHD, did not change the activity level of these same brain areas.[2]

There is no comparable study for learning disabilities, but one can imagine that the brain might be slowed down also. ADHD and learning disabilities often go together.

How does decreased activity in the brain lead to "hyperactivity," inattention and difficulty with concentration? The study showed that the parts of the brain which may be easiest to slow down are the parts of the brain which do the thinking. The parts of the brain which control motor activity are not as slowed down. So the ADHD child or adult has an imbalance between the more active and less active parts of the brain. The parts of the brain that are functioning normally, that is, not slowed down, appear to be more active when other parts of the brain are slowed down. Because the motor activity area is not slowed down, and the thought process area is slowed down, the person appears to be "hyper."

How does the brain come to be slowed down? There are no definite answers. However, as I discussed in Chapter 2, many chemicals both from the diet and from yeast in the intestinal tract sedate or slow down the brain. The chemicals include ethyl acetate from vinegar, pyrazines from malt and alcohols and hydrogen sulfide from internal intestinal yeast. Intake of such chemicals could account for symptoms of ADHD and the research findings I just discussed.

The recognition of ADD/ADHD as a problem follows closely on the heels of introducing antibiotics into standard medical practice. The first major academic study recognizing the problem of attention deficit disorder was published in 1953.[3] The first major antibiotic was introduced into medical practice in the mid to late 1930's.[4] I do not believe this is coincidental.

I suggest, based on my clinical experience and my review of the research literature, that these chemicals may at least be partially responsible for the research findings.

I have treated many patients who suffered from ADHD/ADD. As a child psychiatrist, I have written many prescriptions for Ritalin, Adderall, and other drugs commonly used to "treat" ADD/ADHD. All of these drugs have side effects. None of them alter the basic biology underlying the disorder. Nor is there any evidence that giving these drugs helps a child retain information and learn in the long run.

I have prescribed the 4 Stages diet and nystatin for other patients whose parents were interested in trying a non-toxic therapy rather than conventional psychoactive medications. The success was quick and relatively easy, even for some of my patients who only followed Stage I of the diet. I greatly prefer to alter the basic biology of the disorder by treating yeast, than to continually write prescriptions for psychoactive drugs that just control the symptoms.

All children I have seen who have adopted the 4 Stages diet have calmed down and behaved better. Usually with this diet and the anti-yeast medicine nystatin, drugs such as Ritalin are no longer necessary. Here are some of my cases:

Cases of Patients with ADD/ADHD

Mark

Mark, 15, had a long history of attention problems. He had been on a number of medications. On Dexedrine, he had not eaten very well. Because he was not growing well, the Dexedrine had been stopped. Mark had then grown substantially, but he was still having difficulty with paying attention and with being too active. He was fidgety and restless and could not focus. He was distractible. Mark had no learning problems, but he did not like to do homework. The prior year he had not done homework. He had so much difficulty with concentration he could not concentrate well enough to make Kool-Aid. There were also numerous marital problems in this family. The parents were separated. Mark also had respiratory allergies.

Mark started the 4 Stages diet and nystatin. He came back three weeks later. His parents reported that he was calmer and required fewer reminders to do his homework. He was taking care of his homework on his own. He was now working two nights per week at a convenience store. He was getting along better with his sister. The school had no complaints. He was not dozing off or daydreaming. He was staying on task. He was not wandering off and he was staying with household tasks. His father noted a big change. He had received a B+ on a business test. He was cooperating with wearing his glasses.

In three weeks, years of attention problems and hyperactivity were reversed.

I saw Mark a year later. He has continued to do well.

Anna

Anna is a six and a half year old girl who came to my office because she had some symptoms of ADHD as well as word finding problems and reading problems. I saw her about two months before the end of first grade. Although Anna was tested and found to have average IQ, she was well behind in learning to read in the first grade. She could not sustain mental effort when reading. Her ability to concentrate was limited. She said she became tired when trying to read. She would frequently get out of her chair when she was supposed to be working independently. She also had great social difficulty. Other children could not relate well to her. She was defiant. She was doing strange things, such as making animal sounds and barking. She would poke and pinch others when she was not included in their games. Her mother was concerned about her language development.

Anna had respiratory allergies. Her parents were divorced. When I talked with her, she frequently said I don't know or I don't remember.

She was first diagnosed as having a learning disability due to her difficulty in learning to read as well as having some symptoms of ADHD.

She started the 4 Stages diet. She came in a month later. Anna's parent reported that she was less sleepy after lunch and less tired after reading. Anna was following the rules better. She was making connections more quickly and was understanding better. Her mother said that she would like to see Anna's concentration improve.

When I talked with her, Anna seemed more connected. She volunteered answers to questions. Her eye contact had improved.

She started nystatin in addition to the 4 Stages diet.

Four months later, a month after Anna began second grade, Anna's reading had improved considerably. She was reading nearly at grade level. Her other academic skills were pretty good. She was making some friends. Her behavior at both of her parents' homes had improved.

Two months after that, Anna's teacher wrote that Anna had tremendous academic potential, was working hard at almost all academic tasks and seemed to really enjoy learning. Another teacher wrote that Anna had progressed from an S- to an S+ in reading from first grade to second grade. She was doing outstanding work in spelling.

Anna had brought her schoolwork up markedly in seven months. Her behavior had improved. Overall, Anna was doing much better on anti-yeast treatment.

Dennis

Dennis, 8, was adopted at 18 months from a Russian orphanage. He was first observed to be hyperactive in day care at age 2 and a half. He was still hyperactive. He also rocked in front of the TV and jumped up and down. He had been diagnosed with ADHD at age 6. Dennis had been tried on a number of medications for hyperactivity. Adderall helped him stay in his seat, but he made weird facial gestures. He would moo when writing. He had been tried on the Feingold diet, which may have helped a little. When I first saw Dennis, his activity level had come down some, but he could become obsessed with certain subjects such as clocks and times. On neuropsychological testing he had an IQ of 76. He did better on the non-verbal parts of the test and on the achievement tests. He had difficulty with abstract reasoning. He was still making inappropriate facial gestures. The Adderall had been stopped about a month prior to this appointment.

Dennis had a history of cleft palate, for which he had had two surgeries. He was not walking at 18 months in the orphanage but started to walk quickly after coming to the US.

I prescribed for Dennis the 4 Stages diet and nystatin. Seven weeks later, his mother noted that he was no longer hyperactive and he could sit much better, for longer periods of time. He was not rocking and bouncing as much. He was more focused, for example, on bicycle rides, and when working with the computer on educational programs When he had first been taken off the Adderall, he was "all over". He now was paying more attention to games. He was good in a new day care situation. He no longer needed the Adderall. His sleep was good and his eating was great. His tactile need to fiddle with his shirt was better. His vivid facial gestures had diminished.

In my office, Dennis' activity level was appropriate. His mother said she was pleased with his current behavior.

Melanie

Melanie was a seven year old girl who was impulsive, with low attention span and a changeable mood. Her mother said that Melanie could focus when she wanted to. She was reading at grade level and was so far doing all right. She was good in art and with colors. On testing, she was found to be high in math, but to have low spatial skills. Melanie had little tolerance for friends. She was a leader and was outgoing but bossy. She had trouble keeping friends because she demanded that others do it "her way". She had low tolerance for give and take.

Her mother stated that Melanie could turn into a devil child. If held, she could bite. A request for her to do something, especially if she was hungry or tired, could set her off. She could become angry at transition times, angry at whoever was there. She could be very nice to younger kids. She loved music and dancing. Melanie liked the color blue and colors in general. Her mother stated that Melanie tried to be good, but then something would switch and she would become another person.

Melanie was adopted at birth. Her biological mother was a single mother who drank during the first four months of pregnancy because she thought she would have an abortion. She then changed her mind. The biological father was in jail for aggravated assault.

Melanie's only medical problem was warts on her fingers. She was otherwise healthy and had a high tolerance for pain. She reached her developmental milestones on time and was now in the fourth grade.

In my office, Melanie was restless and was stretching in her seat.

Melanie started the 4 Stages diet and was given the homeopathic remedy Tarentula hispanica 30C.

She came back five weeks later. Her mother noted that she was interacting all right with friends. She had not bitten in the last month. She was calmer and not as restless. She was having good days and bad days, but she had more good days than bad days and her good days were better than her previous good days. Her mother thought that her bad days may be due to eating the wrong foods. Cutting out chocolate in the first week had helped. When Melanie had yeast bread, she reacted badly to it. Her teachers had noted an improvement. Melanie's attention span was still limited but if she was directed to the work, she could stay with it.

In my office, Melanie was calmer and sat most of the time.

At this time, Melanie was started on nystatin. The potency of the homeopathic remedy was increased to Tarentula hispanica 200C and the mother was encouraged to do more of the anti-yeast diet, including removing yeast bread.

Three months after the initial visit, Melanie's mother reported that Melanie was now much more focused. When she became upset, she came out of it faster. She was having mostly good days and she was calmer. The teachers said Melanie was doing better and could stay with work longer. Melanie was learning and she was reading well. She was in a regular class. She still had some outbursts. She was eating more.

Melanie had taken the remedy twice, three weeks apart, and she was taking one fourth teaspoon of nystatin powder four times a day in addition to following the diet. In my office she was smiling and was cooperative.

Notes

[1]Zametkin, A. J., Nordahl, T.E., Gross, M., King, A.C. Semple, W. F., Rumsey, J., Hamburger, S., and R. M. Cohen. Cerebral glucose metabolism in adults with hyperactivity of childhood onset. *New England Journal of Medicine*, 323(2):1361-6, 1991.

[2] Matochik, J.A., Liebenauer, L.L., King, A.C., Szymanski, H.V., Cohen, R.M., and Zametkin, A.J. Cerebral glucose metabolism in adults with attention deficit hyperactivity disorder after chronic stimulant treatment. *American Journal of Psychiatry*. 151(5): 658-64, 1994.

[3]La Pouse, R., and Monk, M., "An epidemioligcal study of behavioral characteristics in children," *Am. J. of Public Health* 48:1134-44, 1953.

[4] Mandell, G.L. and Sande, M.A. Antimicrobial Agents: Sulfonamides, Trimethoprij-Sulfamethoxazole, and Urinary Tract Antiseptics, in Goodman and gilman, *The Pharmacological Basis of Therapeutics, 6th ed.* (1980), 1106.

Chapter 6

Depression and Yeast

Depression improves when a person is treated for intestinal Candida yeast.

First, what is depression?

Psychiatrists define depression as a group of symptoms including down or sad mood, loss of pleasure in usual activities, loss of energy, decreased appetite, poor sleep, and diminished self esteem.   Depression can disrupt and ruin people's lives by taking away their energy and ability to function. Antidepressant drugs help about 75% of people, but cause significant side effects.

No one knows what causes depression.

As a psychiatrist, I have treated hundreds of patients for depression using conventional medications. I have seen what works and what does not, and what the side effects are. I have treated many depressed patients who have been receptive to nontraditional treatment. I can say with certainty that depressed people who are treated with the 4 Stages diet and nystatin improve significantly with none of the side effects of the conventional medications.

I explain this by saying that yeast chemicals slow down the brain and make the brain function less well than it should. Many Americans eat foods which contain chemicals which slow down the brain. The intestinal yeast Candida albicans also makes chemicals which slow down the brain. I discuss these chemicals in detail in Chapter 2. Taking in these chemicals will diminish with anti-yeast treatment. From a clinical perspective, I see the relationship between reducing intake of such chemicals and improvement of depressive symptoms.

I offer the following thoughts on why this may occur. The brain may slow itself down as a reaction and as a protective mechanism against such chemicals. As an analogy, if one is outside breathing toxic chemicals, if one is less active and is breathing less, then one will inhale fewer toxic chemicals. If one is exercising vigorously, one will be breathing harder and will take in more of these airborne toxic chemicals. A similar mechanism may occur in the brain. If the brain is more active, then if toxic chemicals are present, more will enter the brain and affect the brain. If the brain is less active, such toxic chemicals may have less impact.

Although I cannot say for sure why the anti-yeast treatment works, we can see from the cases that it does.

Cases of Depression

Mary

Mary, 37, reported she was tired and fatigued. She also told me she had weakness in her arms when writing and feelings of being clumsy. She was depressed. She had been treated for depression before. She complained of memory problems and of becoming depressed before her period. She slept all the time. She had cut back her work schedule. Mary was having vaginal yeast infections often, the most recent a few weeks before. She also complained of rectal itch, irritability and morning tiredness. Mary had nausea, gas and watery nose, worse in fall. She said that she ate no sugar, but still craved it. She had taken many antibiotics in the past including tetracycline for acne.

I started Mary on Stage I of the 4 Stages diet, and prescribed nystatin.

Mary returned six weeks later. Mary's depression had cleared up completely, as did her nausea. Mary reported that she had only minimal feelings of tiredness in the morning and no clumsy feelings. Her writing was easier with no weakness. The rectal itch was gone except when she was off the diet. Mary no longer needed as much sleep. Her watery nose had improved. She was taking one quarter teaspoon of nystatin three times a day.

Judy

Judy, 24, told me that she had morning nausea, blackouts (seeing black, then white stars for two seconds without falling) one to three times per week. She also told me that she was depressed and crying. Judy had been hospitalized recently for depression.

In addition to these symptoms, Judy told me that she could not control her urine. She was fatigued and had nearly daily headaches in the back of her eyes. Judy suffered from mood swings, and arthritis in her fingers, elbows, shoulders, with occasional leg cramps. She craved sugar. She had gained 14 pounds since quitting smoking eight months previously. She also had acne in the last few months and her hair was thinning. After the birth of her first child four and half years previously, Judy had weak hands. Before the pregnancy, she had been fine. Judy had her first vaginal yeast infection during pregnancy. She also had a right ovarian cyst.

She had been on the oral contraceptive pill for four years, but then her depression worsened. She had severe menstrual cramps with much bleeding. She was forgetful and could not concentrate. She had recently been hospitalized for depression. She had had antibiotics in the past. She thought she might have a vaginal yeast infection at the time of the appointment.

Judy started the 4 Stages diet and nystatin.

She came back three weeks later. She told me that she was feeling good. Her blackouts were gone and the morning nausea was much better. Her emotional state was good; her energy was better and her crying was gone. Her ability to control her urine was better. She had had headaches only when she increased the nystatin dose. Her mood swings were better. Her acne was better and her hair felt thicker. Her arthritis was better but not gone. She craved sweets premenstrually, not all the time. She had had a vaginal yeast infection which cleared. Her concentration was about the same. She was taking one half teaspoon of nystatin four times a day.

Melissa

Melissa came in at age 48. She was very depressed and felt like she was going to die. Melissa felt that she was obese, "to the point where I'm dying." She was an insulin dependent diabetic and had trouble sleeping. She had had panic attacks for one month. She stated that she was up and down all night. Melissa had a bad back, and muscle spasms at night She had vaginal yeast infections since using antibiotics. She had alternating diarrhea and constipation and a poor energy level, occasional headaches, pains in the nerves in her legs with swelling. She had a history of bad ear infections and she had used many antibiotics.

After two months on the 4 Stages diet and nystatin, Melissa was much better emotionally. She had lost weight and had much more energy. Her appetite was way down. Her yeast infections were gone. She had been able to reduce her insulin dose. The muscle cramps in her thighs were gone. Significantly, Melissa's food cravings were much less intense and did not last as long. She still had panic attacks but she could get through them.

Julie

Julie, 34, told me she was very depressed. She had been through divorce and bankruptcy in the last two years. Five years previously she had been depressed when marital problems had begun. Julie's depression worsened a few months previously after the breakup of a second relationship. She had a vaginal yeast infection, for which she had used Monistat and Flagyl (an antibiotic) which had helped, but she still had some vaginal discharge. But a few weeks after being on the Flagyl, Julie had become more depressed.

One of Julie's main concerns was that she "overdoses" on chocolate. This was a long standing problem. She had been very obese weighing 310 pounds at one point. She lost weight by having the drastic gastric bypass surgery. Her coworkers and boss had complained of her mood swings and she had been suicidal in the past. She had a hysterectomy the year before for excessive bleeding. She had had many antibiotics in the past including a heavy course for a postsurgical complication.

I treated Julie with the 4 Stages diet and nystatin. Julie completely turned around in five weeks. Julie felt excellent! She reported that she had started to feel better five days after starting the 4 Stages diet and nystatin. Her vaginal yeast infection was gone and she had stopped eating chocolate.

Chapter 7

Food Cravings, Addictions and Yeast Chemicals

Many patients suffer from feelings that they are not in control of their eating. This problem may be termed food addiction.

To understand how to break the food addiction cycle, you need to understand why you get addicted to food.

Let's start with definitions. Food cravings are extremely strong desires for certain foods. These cravings are so strong that you focus on the craving above everything else. And you focus on food when you no longer need it to sustain you. Addiction results when you lose control and give in to these intense desires regardless of consequences.

I see many patients in my medical practice who are addicted to food, and I help them regain control.

My patients, and people with whom I talk, are most commonly addicted to three types of foods: sugary foods, chocolate, and snack foods that are high in fat and salt, like potato chips. I have heard a lot from "chocoholics," but have never heard anyone describe themselves as a "grapefruitholic". Patients never describe addiction to foods like lettuce, grapefruit, blueberries, or even meat.

What does this tell us? Food addiction is not about the need for sustenance. Many calorie sources would provide sustenance, but the average food addict would never touch them to satisfy the addiction. So something is unique to the addictive foods that goes well beyond the body's need for survival.

When we think of chocoholics or food addicts, we think of the bingers—people who can eat huge quantities of food in one sitting. But addiction is not just about eating a gallon of ice cream. Addiction is also about not being able to finish the day without an extra sweet dessert. It is about thinking about chocolate all the time, and searching for that old candy bar at the bottom of a purse. Medical science does not seem to understand these intense urges and addictions.

Why do I call people who crave food "food addicts"? The definition of addiction includes physiological dependency on a substance even when there are significant negative consequences to using or taking in that substance. Despite knowing the consequences, the addict must ingest the substance.

The main obvious consequences of food addiction are weight gain and loss of self control. Hidden consequences include depression, medical problems, loss of energy, and other major problems. Food addiction is legal but its cost is not trivial. Millions of Americans are on weight loss diets at any given time. The weight loss industry is a multimillion dollar industry, but diets and pills do not solve the underlying problem of irresistible cravings. That's because food addiction has a physiological basis and cannot be stopped by sheer will power.

What's behind the cravings? What is the physiological basis of food addiction?

Food addiction is based in biology, not psychology. No amount of will power will eliminate cravings and addictions. To understand why, you need to understand a little about food chemistry.

Many chemicals in our food affect our brains and can trigger addictive behavior. Unfortunately, the established medical research community basically has never studied how the chemicals in food affect our eating behavior. This lack of medical research truly is tragic. Despite this research gap, my own research, clinical experience with numerous patients, and the data that does exist show that what we eat has a profound affect on every aspect of our lives.

I've wondered about food cravings ever since I learned about bulimia as a medical student. Patients with bulimia both binge on large amounts of high calorie foods and then try to purge, usually by vomiting but also by abuse of diuretics (to urinate more) or laxatives. I wondered why bulimics binged on ice cream and not on grapefruit. This question stayed with me through my doctoral studies in nutrition at the University of California-Davis, and later in my research on the role of diet in cancer at the National Cancer Institute. My questions went unanswered. Even my training as a psychiatrist did not help answer this simple question.

A new approach: the 4 Stages Diet

Ultimately, my practice as a medical nutritionist and psychiatrist and my many hours of research helped me understand food cravings better than did any formal training. As a child psychiatrist and as the father of a child with autism, I was trying to help children with autism and other similar developmental problems. After examining all the available research about what in food affects the brain, I concluded that by changing their diets, children with autism could be treated more effectively than by drugs. As a result, I developed a special diet to treat such children.

The results of my research on autism are described in Chapter 4. Children who spoke only one or two words developed language; children who could make no eye contact were connected. Balance improved and these children could function more normally in the world. I called the diet the "4 Stages" diet, based on four different levels of need. Interestingly, parents would describe their children's addictive behavior. Almost all of the children craved foods such as ketchup and chocolate. After being on the 4 Stages diet for a while, the parents reported that virtually all the cravings disappeared.

As described in other chapters, I expanded to prescribing the 4 Stages diet for patients who suffered with symptoms from illnesses ranging from depression to chronic fatigue to fibromyalgia. The results were the same. These patients also reported food cravings at the beginning of treatment. After time on the diet, however, their food cravings improved significantly, and many lost weight.

How it works

So, what was happening that the same food changes that helped autistic children function and helped chronically fatigued people regain energy, also eliminated food cravings? We've always been told that food cravings were psychological. Could food cravings and addictions be treated by changing diet? Certainly.

We can look again at Carol's case, which I presented to you in Chapter 3 when discussing headaches. Carol, like many patients, suffered from multiple problems relating to yeast. Two of her main problems were headaches and chocolate cravings.

Carol

Carol, 43, came to see me because she had suffered from premenstrual syndrome, extreme headaches, a plugged up nose and breathing troubles. This had gone on for the past 6 years. She had a hiatal hernia, so she had been told to avoid coffee and chocolate before her period. Carol had intense chocolate craving during this time. Allergy testing showed a few fall allergies. The sinus problems were so bad she had sinus surgery the year before. When Carol came to me she was getting nosebleeds for a week before her period. Her nose was swelling up, especially at the incision sites. She still had headaches. She had a stressful job and could not concentrate. Concentration was worse before her period. She also had bloating and gas premenstrually.

Carol had patches of dry skin on her scalp and face. Her skin itched before her period. She had no nasal congestion or headaches at other times of the month. Her periods were only 21 -28 days apart, so she experienced these symptoms every few weeks. Carol had taken many antibiotics the year before the problem started. She had three children.

After discussing her problems and symptoms, I prescribed the 4 Stages diet for Carol. Two and a half months after being on the diet and taking nystatin, Carol came back feeling "pretty good". Her chocolate cravings had greatly diminished. Her headaches had decreased from being debilitating for days at a time, to only the day before her period. They were much less intense. She had more energy. Her nose was still plugging up and bled some, but less. The itching on her head was gone. The dry patches of skin on her ears and eyebrows were gone but the dry patch on her scalp was still present. Her bloating was gone.

What in chocolate causes craving?

To understand how the diet worked to eliminate Carol's chocolate addiction, we need to know what is in chocolate. Chocolate contains a chemical called tetramethyl pyrazine, a very important chemical in the flavor of chocolate.[1] Chocolate also contains several other pyrazines.[2] Although researchers found that tetramethyl pyrazine is a very important chemical in the flavor of chocolate, why is this chemical important in food addiction also?

Tetramethyl pyrazine and other pyrazines are very active in the brain. These chemicals slow and sedate the brain.[3] In large enough quantities, these chemicals would put the brain to sleep. Chemicals that put the brain to sleep can be considered in the group of chemicals we call anesthetic agents.

What do such sedating chemicals do?

Such chemicals cause release of brain chemicals called endorphins.[4] Because endorphin release is thought to be pleasurable, eating chocolate may activate a pleasure response in the brain. The more you eat, the better you feel, and not just because chocolate tastes good. But the release of endorphins and the pleasure is temporary, so you need to keep eating the chocolate to keep up your pleasure. People become addicted to chocolate. A little bit doesn't satisfy them. Instead, they need more and more, and they keep seeking it out. Although we may joke about the "chocohalic", this type of addiction is not funny to the people who have it. They are out of control.

What will happen if you try to go cold turkey off chocolate?

So, the usual response is to attempt will power to avoid chocolate. This never works in the long run. Why not? Your body has learned what chocolate does. It releases endorphins which leads to a temporary pleasure response. Your body remembers this feeling and wants to bring it back. The desire and the craving will remain.

Other foods you are eating also keep your chocolate cravings alive.

The same chemicals in chocolate, tetramethyl pyrazine, and other pyrazines, are found in other common foods, so even if you try to exercise will power over chocolate, other foods will sabotage you. They keep on activating this same pleasure center. And as long as you activate it, you'll want to keep on activating it.

But, to quote a cliche, what's so bad about feeling good? Nothing, as long as the costs are not so great. But many people suffer serious adverse side effects. The foods that contain the chemicals that make you feel good also have other effects. Chocolate is high in fat and calories and low in fiber so most chocolate addicts gain weight. Chocolate also causes headaches. Some people become depressed over the weight gain. They want to feel good, so they eat more chocolate, which feeds the addiction cycle.

So now we see why people binge on chocolate, but not on green beans or grapefruit. Green beans and grapefruit do not have the same chemical content of chocolate, namely pyrazines, that triggers our pleasure centers.

What else triggers food addiction?

The food that most commonly comes to mind is sugar. Books have been written about sugar, and many people try to eliminate it. Do people crave just sweet things? Let's look more closely.

Many foods are sweet, but some are addictive and others are not. One of the sweetest natural foods known is honey. Honey is not addictive. Honey is actually hard to eat in large quantities. Nobody has ever come to me for help complaining that they eat honey out of the jar. So pure sweet taste is not the only issue here.

Even when people talk about sugar cravings, they usually do not mean that they take sugar out of the sugar bowl. Usually they mean that they eat sugary foods, often baked goods such as cookies, cakes, doughnuts, sweet rolls and the like. Later, we'll discuss ice cream. What else is in these baked goods that causes addiction?

The Hidden Enemy - Malt

The most dangerous food for food addicts is "malt". Almost all commercial, and many home-made baked goods, contain this sweet substance. Malt comes from barley, which is specially raised to be made into beer. Under special conditions, the barley is sprouted and the starch in the barley is broken down into sugar chains of varying sizes. Then the sprouting barley is heated. Many chemicals form during this process. After heating, the sprouted barley is called malt. Malt goes by a number of names, such as malt, malt syrup, malt extract, and barley malt. Maltodextrin is malt mixed with something else, usually corn syrup because corn syrup contains dextrose, a form of sugar.

Malt is mostly raised to form the raw material for making beer, but it is cheap, sweet and is sold as a sugar substitute, which is then baked into nearly everything, from cookies and crackers to bagels and breads. Malt flakes are in many common brands of white flour, so you don't escape malt by baking your own cookies.

As the barley is sprouted and heated and becomes malt, twenty pyrazines form.[5] Does this word sound familiar? Some of the pyrazines found in malt, including tetramethyl pyrazine, are the same chemicals found in chocolate. When pyrazines are eaten, they are going to release endorphins. If you really want to feed a food addiction, try some malt. Malt is the most addictive part of sweet baked goods.

How does sugar hurt you?

Sugar has its own problems too. To understand how sugar causes food addictions, we need to understand what is in the human intestinal tract. In the intestine, there are microorganisms, the yeast Candida albicans, and bacteria. As described earlier, Candida is a yeast that inhabits the intestinal tract from the mouth on down. This yeast is also present in the vagina. If yeast invades the rest of the body, the infection can be lethal. This type of yeast infection often occurs in cancer patients when immune systems are compromised.

For people with intact immune systems, Candida stays in the gut. Bacteria are much smaller than the yeast Candida albicans. Yeast resemble our own cells and are about the same size as other body cells. After people take antibiotics, yeast are found in larger quantities in the intestine because antibiotics kill bacteria and make room for the yeast to grow.[6] I have found in my clinical experience that almost everybody with food addictions has taken antibiotics, so they almost all have some yeast in the intestinal tract. What do yeast do? Commercially, yeast is used to generate by the process of fermentation, alcoholic beverages, like wine and beer.

In our intestinal tracts, yeast also ferment sugar. Sugar goes into the intestine and mixes with the yeast, and just as in commercial factories, the yeast transforms the sugar into alcohols.[7] What does alcohol do? Alcohol causes endorphin release.[8] Remember, endorphins make us feel good. This is the same process as in chocolate addiction. So people become addicted to sugar, because sugar leads to increased yeast activity, which leads to alcohols and endorphin release. Eating pure sugar provides some food for the body, but pure sugar will do more than this; it feeds the yeast.

Experiments show that the yeast Candida loves sugar and becomes very active after getting it. Then the yeast makes even more chemicals, which continues the cycle. Kids (and adults) can be addicted to soda, which is almost pure sugar. The addiction to baked goods is usually a combination of addiction to sugar and to malt.

So if you try to conquer your craving for sugar by eliminating sugar, you won't succeed. Many new health food snack items in the store labeled "contains no sugar" use malt as a sweetener. Malt may also be called a "grain-based sweetener."

Malt also contains growth factors for the yeast.[9] So eating malt feeds the yeast, which make alcohols, and which will cause endorphin release. Malt contains the chemicals pyrazines, which may also cause endorphin release.

By avoiding only sugar, you will not be taking steps to successfully conquer your sugar craving. The malt still feeds the addiction; the malt feeds the yeast and malt is still causing the endorphin release. The food is still controlling you.

Other addictive foods

There are other addictive foods, including common foods like potato chips, coffee and roasted peanuts. Let us take a look at them.

Potato chips are heated in oil at high temperature. These are the conditions for generating pyrazines and potato chips contain pyrazines.[10] Roasted peanuts and coffee also contain pyrazines.[11] Remember, pyrazines are the same chemicals in malt and chocolate that cause addiction. This is why potato chips and roasted peanuts are so hard to stop eating and is one of the reasons why coffee is so hard to stop drinking.

Ice cream and other foods containing vanilla are also addictive. Vanilla contains its own toxic chemicals.[12] The chemicals in vanilla have some of the same effects as the pyrazines.[13] Both are sedative chemicals. Vanillin may release endorphins also, which would lead to food addiction.

The cases show that the normal tight regulation of eating can be restored by avoiding foods which contain chemicals which activate brain endorphin release.

Here are two other cases where changing what the patients ate solved their food addiction problems.

Solving Food Addiction: Two Cases

You may remember Melissa and Julie from Chapter 6 on depression. Both Melissa and Julie were depressed and had bad food cravings. Often these problems go together. By following the 4 Stages diet, both women solved their food cravings as well as their depression. Here is what happened to Melissa:

Melissa

At 48 years old, Melissa had such bad food cravings she could eat a "red dog running." She felt — and was— obese, "to the point where I'm dying," she told me, her doctor. Melissa was an insulin dependent diabetic. She had panic attacks, trouble sleeping, and was up and down all night. Melissa was very depressed, so depressed she felt like she was going to die. She also had many physical problems, including a bad back, and muscle spasms at night. Melissa suffered from alternating diarrhea and constipation, as well as vaginal yeast infections. Her energy was low. Her nerves in her legs were painful, and her legs were swollen. In short, when Melissa came to me she felt terrible.

Melissa desperately wanted to get better. She had tried everything-psychotherapy, dieting, "12 step" programs. Nothing had helped.

Melissa told me that only sheer force of will kept her alive.

After only two months on my program, Melissa completely turned around. Her life became livable. What did this? I started Melissa on a new approach, a special dietary program that removed foods triggering food addition.

I determined that Melissa's real problem was that she was addicted to food. She felt miserable because she was eating the wrong foods, and not eating the right foods. The very foods she craved were failing to satisfy her. They made Melissa crave more and more. They also made Melissa gain weight to the point of obesity; she became depressed, and they contributed to her serious medical problems.

I prescribed the 4 Stages diet, not based on calories and willpower, but based on eliminating food addiction. I also prescribed a very safe, nontoxic medication, nystatin, that killed the yeast inside Melissa.

The results were amazing. After only two months, Melissa's emotional state had improved dramatically. She had lost weight and had much more energy. Her appetite was way down. Gone were her yeast infections and muscle cramping in her legs. She had been able to reduce her insulin dose. Melissa's food cravings were way down, too. When she had cravings, they did not last as long. The panic attacks were less frequent and not as severe.

Julie

Julie also suffered from multiple problems. Julie came in when she was 34 years old. She was very depressed. She had been through divorce and bankruptcy in the last two years. Five years previously she had been depressed when marital problems had begun. Julie's depression worsened a few months before she saw me. This followed a relationship breaking up. Julie also suffered from a vaginal yeast infection, for which she had used Monistat and Flagyl (an antibiotic). These medications had not completely eliminated the infection. A few weeks after being on Flagyl, Julie became depressed.

One of Julie's long standing concerns was that she "overdoses" on chocolate. She had been very obese, weighing 310 pounds at one point. She lost weight by having the drastic gastric bypass surgery. Her coworkers and boss had complained of her mood swings and she had been suicidal in the past. She had a hysterectomy the year before for excessive bleeding. She had had many antibiotics in the past including a heavy course for a postsurgical complication.

I treated Julie with the 4 Stages diet and nystatin. Julie completely turned around in five weeks. Julie felt great! She reported that she had started to feel better five days after starting the special diet and nystatin. Her vaginal yeast infection was gone and she had stopped eating chocolate.

Why did these patients improve?

These patients improved because they initially avoided foods which triggered their food addiction, like chocolate. They added foods that helped their bodies, and they also took nystatin, a non-toxic anti-yeast medication. Avoiding addictive foods gave their bodies and brains a chance to heal and re-regulate themselves. The 4 Stages diet works in a positive way. When these patients avoided these foods, they also did not feed the yeast in their intestinal tracts. Then the yeast became less active and made fewer chemicals. Their brains worked better. Their eating centers re-regulated. Julie, Melissa and Carol and my other patients ate less of what was bad for their bodies and more of what was good. They got rid of their depression, fatigue and yeast infections. Most important, they got rid of their food addictions. So they no longer wanted the chocolate or chips or whatever fed their addictions. They could pass by chocolate and say "no, thanks," without feeling that tug of addiction. The vicious cycle was turned off.

What is the answer to food cravings and food addictions?

The answer is first to know that there is a biological basis for food addiction and that one can learn how to avoid the foods which trigger this biological addiction. Following the appropriate food choices as given here and in our cookbooks *Feast Without Yeast* and *Extraordinary Foods for the Everyday Kitchen* will help you.[14] You will learn how to gradually eliminate the foods you crave and to substitute better foods. You will start to feel better and more in control. Most people lose weight as they regain control over their eating. You do not have to be a food addict. You do not have to be a slave to ice cream or chocolate. You will learn how to recognize what food trigger your addiction, and how to eat those foods in a controlled way.

Notes

[1] Zak, D. L., Ostovar, K. and V. G. Keeney. Implication of Bacillus subtilis in the synthesis of tetramethylpyrazine during fermentation of cocoa beans. *J. Food Sci.* 37(1972), 967-68.

[2] Rizzi, G. P. The occurrence of simple alkylpyrazines in cocoa butter. *J. Agr. Food Chem.* 15:549-551, 1967.

3 Nishie, K., Waiss, A. C. Jr., and A. C. Keyl. Pharmacology of Alkyl and Hydroxyalkylpyrazines. *Toxicology and Applied Pharmacology.* 17:244-249, 1970.

4 Maiewski, S., Muldoon, S., and Mueller, G.P. Anesthesia and stimulation of pituitary beta-endorphin release in rats. *Proceedings of the Society for Experimental Biology and Medicine.* 176(3):268-75, 1984.

5Malt contains chemicals called alkyl pyrazines (Hough, et. al., 1982). Hough, J. S., Briggs, D. E., Stevens, R. and T. W. Young. Brewery Fermentations. *Malting and Brewing Science,* 2nd ed., Chapman and Hall, London, vol. 2, pp. 462-471, 1982.

6 Samonis, G., Anaissie, E.J. and G.P. Bodey. Effects of broad spectrum antimicrobial agents on yeast colonization of the gastrointestinal tracts of mice. *Antimicrobial Agents and Chemotherapy.* 34:2420-2422, 1990.

7Yeast make many alcohols, both ethyl alcohol and other much more poisonous alcohols such as 1-propanol, 2-propanol, 1-butanol and 2butanol and phenyl-ethyl alcohol (Rankine, 1967, Lingappa, et al, 1969 and Berry and Watson, 1987). Lingappa, B.T.., Prasad, M., Lingappa, Y., Hunt, D.F.., and K. Biemann. Phenethyl Alcohol and Tryptophol: Autoantibiotics Produced by Fungus Candida albicans. *Science.* 163:192-193, 1969. Rankine, B.C. Formation of higher alcohols by wine yeasts. *Journal of the Science of Food and Agriculture.* 18:583-589, 1967.Berry, DR., and D.C. Watson. Production of organoleptic compounds. In *Yeast Biotechnology,* Berry, DR., Russell, I and Stewart, GG., Eds., London, Allen and Unwin, 1987, p. 345.

8 Seizinger, BR., Bovermann, K, Hollt, V. And A. Herz. Enhanced activity of the Beta endorphinergic system in the anterior and neurointermediate lobe of the rat pituitary after chronic treatment with ethanol liquid diet. *The Journal of Pharmacology and experimental Therapeutics.* 230:455-461, 1984.

9 Croft, C. C. and L. A. Black Biochemical and morphologic methods for the isolation and identification of yeast-like fungi. *J. Lab. clin. Med.,* 23, 1248-58, 1938.

10 Deck, R.E., and Chang, S. S. Identification of 2,5-dimethylpyrazine in the volatile flavour compounds of potato chips. *Chem. Ind.* No. 30, 1343-1344, 1965.

11 Mason, M. E., Johnson, B., and Hamming, M. Flavor components of roasted peanuts. Some low molecular weight pyrazines and pyrrole. *J. Agr. Food Chem.* 14, 454-460, 1960. Gianturco, V. M., Giammarino, A. C., and Friedel, P. Volatile constituents of coffee. *Nature,* 210, 1358, 1956. Viani, R., Muggler-Chavan, R. D., and Egli, R. H. Sur la composition de l'arome de cafe. *Helv. Chim. Acta.* 48, 1809-1815, 1965.

12Riley, K. A. and D. H. Kleyn. Fundamental Principles of Vanilla? Vanilla Extract Processing and Methods of Detecting Adulteration in Vanilla Extracts. *Food Technology.* 43(10): 64-77, 1989.

13Large amounts of vanilla cause coma.   Jenner, D. M., Hagan, E. C., Taylor, J. M., Cook, E.L. and O. G. Fitzhugh. Food Flavourings and Compounds of Related Structures. I. Acute Oral toxicity. *Food and Cosmetics Toxicology.* 2:327-343, 1964.

14 Semon, B. A. and L. S. Kornblum. *Feast Without Yeast: 4 Stages to Better Health* (1999: Wisconsin Institute of Nutrition, Milwaukee, Wisconsin); Semon, B.A. and L.S. Kornblum, *Extraordinary Foods for the Everyday Kitchen* (2003: Wisconsin Institute of Nutrition, Milwaukee, Wisconsin), 1-877-332-7899, http://www.nutritioninstitute.com.

PART III
How Yeast Affects Our Immune System and Our Health

Chapter 8

How Yeast Evades Our Immune System and Causes Medical Problems

---

The immune system and yeast: a battleground

Every person has a body system, called the immune system, that fights, and usually destroys, infectious foreign invaders. The immune system sees the yeast Candida as a foreign invader to fight. Sometimes, there is no clear winner, because Candida evades the body's immune system. The fight between the immune system and Candida continues. This continued fight can lead to major health problems from multiple sclerosis to ulcerative colitis.

Because the immune system is complex, and because Candida interacts with the immune system in different ways, I have organized my discussion into several chapters.

In this chapter, I will first explain the basics of how the immune system fights Candida, how Candida evades the immune system, and how these fights lead to persistent

153

inflammation and disease. Treatment of the yeast Candida ends these fights and heals these health problems. I will discuss the types of illnesses linked to immune problems in separate chapters. I will discuss medical conditions characterized by chronic inflammation, such as Crohn's disease and ulcerative colitis in Chapter 9. Yeast and skin disorders, including eczema, psoriasis and chronic rashes, are covered in Chapter 10. Medical conditions called autoimmune disorders, including multiple sclerosis, and rheumatoid arthritis are discussed in Chapter 11. How yeast contributes to allergy problems is discussed in Chapter 12. Yeast and recurrent ear and sinus infections are explained in Chapter 13. Yeast and recurrent vaginal yeast infections is discussed in Chapter 14. Yeast and hormonal problems is discussed in Chapter 15. In each chapter, I give case examples showing you how yeast can be treated and how the patient regains health. I will also discuss the scientific studies supporting these findings.

What led me to look at Candida and the immune system? In medical school, and as a practicing doctor, I was curious about and confused by a certain class of illnesses called autoimmune disorders. These disorders include multiple sclerosis, rheumatoid arthritis, psoriasis, ulcerative colitis, Crohn's disease and others.

These are complicated disorders which medical researchers have not yet been able to understand. Generally, it is thought that the immune system goes haywire and attacks the body's own organs. These conditions are chronic, that is, they become part of a person's life. There is considered to be no cure for such disorders. Most medical treatment is designed to relieve symptoms and pain.

How might yeast cause all of these chronic diseases? These diseases all are marked by the body's immune system attacking the body's own organs in what seems to be a prolonged destructive war with no resolution.

I have always considered strange the idea that the body's immune system attacks the body's own organs for no apparent reason. The body's immune system is designed to fight foreign invaders, not our own organs. What could make our immune system instead attack the body's own organs? In 40 years of research, the biomedical research community has been unable to come up with any clear answer.

Maybe the concept is wrong. Maybe the primary target of the immune system is not our own organs. Maybe the primary target really is a foreign invader that scientists have not noticed because it is not found at the attack site. Maybe the body is a secondary target that gets caught in the crossfire as the body's immune system attacks the foreign invader. Medical researchers have considered the possibility that a foreign invader stimulates the immune system to attack the body's own organs. To date, they have not identified such a foreign invader.

To understand how Candida could be this foreign invader we first need to understand something about the immune system. In the rest of this chapter, I will discuss the immune system. The information is this chapter is critical for your understanding of specific medical conditions. How yeast causes autoimmune disorders, as well as case descriptions of treatments of people with such disorders comes in Chapter 11.

How your immune system protects your body from foreign invaders

Foreign "bugs" or microorganisms constantly beseige the human body. These microorganisms enter your body through your mouth in food, water, and air. They also can enter your body through your skin, including through sexual contact. Microorganisms are tiny, and many are smaller than our own cells. Microorganisms include yeast, fungi, bacteria, viruses, one celled organisms such as amoeba, and multi-celled organisms such as protozoa.

Microorganisms can multiply rapidly and kill an animal or human being unless the animal or human mounts a defense. The defense is called the immune response, which the immune system initiates and controls. The immune system is complex and consists of a number of types of cells, with different functions and capabilities, because the microorganisms come in different sizes and have different strategies. All of these immune cells must interact together to produce the best immune response.

For example, one immune response is for certain immune cells to "eat" smaller microorganisms such as bacteria. Bacteria release some chemicals as part of their chemistry and metabolism. These chemicals attract specific immune system cells which then "eat" the bacteria. This process is called phagocytosis.

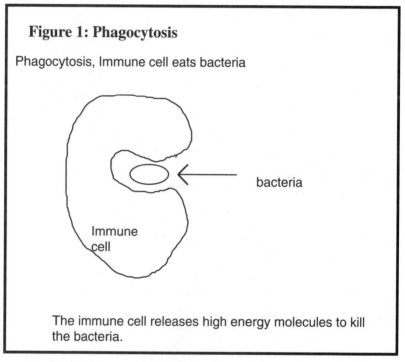

Figure 1: Phagocytosis

Phagocytosis, Immune cell eats bacteria

bacteria

Immune cell

The immune cell releases high energy molecules to kill the bacteria.

Another immune cell technique is to kill infected cells. Viruses attack by entering our own cells and taking over the cells' machinery to make more viruses. The immune system destroys viruses by killing virus-infected cells.

Immune cells have other techniques for responding to foreign invaders. Some microorganisms are larger than bacteria and viruses, such as yeast and fungi. Other types of immune cells respond when they detect a foreign invader such as yeast and fungi, and use other techniques.

Immune cells detect foreign invaders by recognizing that the foreign invader's cellular structure is different than that of any body cell.   All cells put out receptors, which are branch-like structures; among human cells, these receptors enable cells to communicate with one another.   Microorganisms and virus-infected cells have different receptors.   Immune cells can recognize these receptors as foreign, and can respond by attacking the foreign cells.

Figure 2: Immune Cell, Yeast Cells, Virus Infected Cells and Bacteria

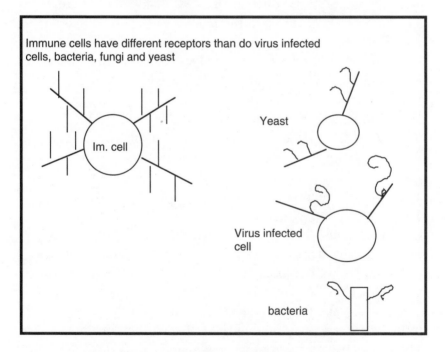

Immune cells work in many different ways.

The techniques of directly contacting and destroying foreign invaders, and of killing virus infected cells are only two of the tools for fighting foreign invaders.

The immune system could not function without signals between immune cells. Some immune cells, called helper cells, direct other cells by making signals. Helper cells can make the signals for other cells to generate antibodies or to generate inflammation. Both antibodies and inflammation will be discussed below.

Another tool to fight foreign invaders is for some immune cells to make antibodies, which are large molecules that attach to infectious microorganisms. When these antibodies attach, other immune cells can "eat" the infectious agents more easily.

Some immune cells generate the signals for another immune system tool, inflammation. When you get a cut, the area around the cut may become red, swollen, hot and painful, all signs of inflammation. Inflammation is a major defense against foreign invaders. Inflammation is a process of increasing blood flow to an infected area with swelling, pain and heat. This process brings in immune cells to fight the foreign invader. Inflammation is painful, but serves a purpose. Inflammation puts a wall between the body and foreign invaders to contain the microorganisms to a small area. Then the problematic microorganisms have a much harder time spreading. How your immune system generates inflammation and how inflammation protects your body, is very complex. The signals of inflammation attract other immune cells to fight the battle against the foreign invader. The swelling prevents the foreign invader from moving too much. Inflammation is a necessary part of fighting foreign invaders, which includes the yeast Candida. Keep this thought in mind as you read further.

To make sure the immune cells do their job and go no further, there are other immune cells, called suppressor cells. These cells make sure, for example, that the inflammation is only generated where it is supposed to be generated. They downregulate (slow and limit) the system.

So now we know the immune system uses many tools to fight the foreign invaders. Generally, those tools are eating foreign invaders (phagocytosis), killing cells infected with foreign invaders, making signals to communicate with other immune cells and to generate inflammation, and making antibodies to attach to foreign invaders

Why is the immune system so complex and diverse? Think of your immune system as your internal defense department consisting of army, navy and air force. Your body does not know what types or numbers of foreign invaders will attack. So your internal defense department cannot rely on just an army, or just an air force or navy. It needs all the tools, weapons and techniques possible to keep you safe. Your body must be able to handle all the different kinds of challenges from so many different microorganisms.

When your immune system detects a foreign invader, immune cells go into action, making signals to direct production of antibodies, to eat microorganisms and to produce inflammation. Even if all the immune system tools are needed, the immune system must destroy the foreign invader! Otherwise you might die from a simple infection.

So now we know that the immune system consists of many parts which are coordinated and use signals to communicate. The system is designed to kill foreign invaders and can generate significant amounts of inflammation.

I like to think about our immune system this way. Just as in the military, your body needs both offensive and defensive weapons. Certain parts of the immune system are the offense. These parts seek out and kill the foreign invaders before they harm you. Other parts of the immune system, such as inflammation, are the defense. These parts protect your body by preventing the invading microorganism from spreading. In a typical infection, we see both parts at work. There is inflammation, which is defensive, but inside the inflamed tissue, the body's immune cells are offensive and are killing the foreign invaders. This concept of both offensive and defensive parts to the immune system is important when we look at how the immune system handles Candida.

Another part of your body's defense system - the good bacteria: commensal microorganisms

We have been talking about what the immune system does when it detects a foreign invader. Our bodies have yet another defense against foreign invaders: good, or commensal bacteria. Infectious microorganisms cannot do anything unless they can stick, or adhere, to a body surface. This concept is very important to keep in mind. As a defense against this adherence, the body uses certain bacteria, called commensal bacteria, to defend against bad, infectious microorganisms. These commensal bacteria occupy places, or sites on body linings, such as in the throat and gut (intestine), and on the skin. These commensal bacteria take up space on these linings, and they make it harder for the bad microorganisms to stick to a lining and multiply.

Figure 3: Good bacteria lines body surfaces

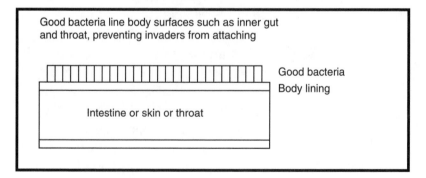

The term commensal is defined as two organisms living together, both helping the other with neither harming the other nor being a parasite of the other. Our intestines have numerous bacteria present in the gut which fit this definition. These bacteria benefit people by adhering to the inner intestinal lining and preventing bad microorganisms from coming in and causing disease. Truly commensal bacteria adhere without causing any inflammatory reaction. We can look at this relationship as good neighbors who watch out for each other and help each other. Neither one causes the other problems, but if a potential criminal comes around, each will call the police to protect the other.

Antibiotics kill commensal bacteria

The reason commensal organisms are so important is they defend your body against bad microorganisms. We should not take this line of defense for granted. The main threat to these commensal bacteria comes from using antibiotics to treat illnesses. When people take antibiotics to kill infectious bacteria, such as for strep throat, the antibiotics do not distinguish between the strep bacteria and the good bacteria. The antibiotics kill all the bacteria, including these commensal organisms. Today's broad spectrum antibiotics such as Augmentin and Ceclor, are better at killing a variety of bacteria than were older antibiotics such as penicillin. However, these newer antibiotics also kill more commensal bacteria. So antibiotics are designed to help you by killing infectious bacteria which also remove this natural defense of commensal organisms.

Figure 4: The same body lining after antibiotics

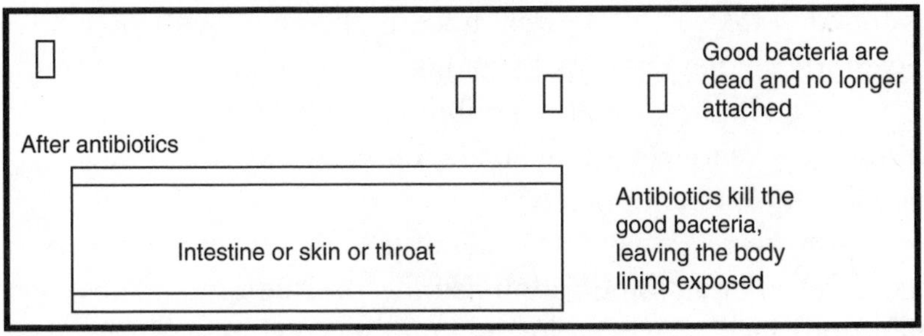

Candida takes root after antibiotics kill the commensal bacteria

What happens on a body surface after antibiotics have killed the defensive layer of commensal microorganisms? The death of commensal microorganisms leaves an open surface for other infectious microorganisms to attach and grow. The yeast Candida is one microorganism which can and does grow on a body surface, especially after antibiotics kill the commensal bacteria.

Figure 5: How yeast cells attach to body linings after the good bacteria are gone

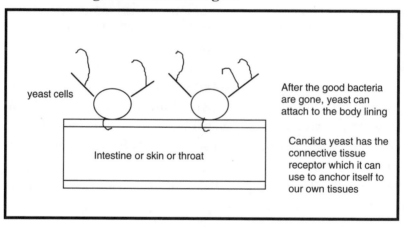

yeast cells

Intestine or skin or throat

After the good bacteria are gone, yeast can attach to the body lining

Candida yeast has the connective tissue receptor which it can use to anchor itself to our own tissues

Even without using antibiotics, doctors know that Candida is found in the gastrointestinal tract, in the mouth and vagina.[1] Scientific studies show that Candida grows at these sites and in the gut at a higher rate after antibiotics have been given. Studies also show that Candida is found on the skin after people take oral antibiotics.[2]

How do we know that Candida stays in the gut and increases after antibiotic administration? Studies also show that yeast increases in the feces after use of antibiotics, [3] showing that there is increased yeast in the gut after taking antibiotics.

Candida does not go away by itself

Once Candida is allowed to grow on body surfaces, and linings, such as in the gut, in the mouth and on the skin, Candida often stays. This is well known. In the past, doctors considered important reporting that lengthy use of antibiotics led to serious Candida infections.[4] Some drug manufacturers used to package nystatin, the anti-yeast drug, together with their antibiotics, because they knew taking antibiotics could lead to serious yeast infections and wanted to prevent them.[5] Even though there are fewer such reports, the problem has not gone away.   Instead, some patients have persistent Candida infection syndromes which are resistant to treatment.[6]

Large amounts of Candida are not normal, and can harm you

Studies show that Candida levels increase after a person takes antibiotics. So what? Is this bad? Yes. Is Candida albicans meant to be present in large quantities in the human gut? No. No studies show Candida does anything positive for people. Candida is capable of causing considerable harm.

Remember that Candida makes numerous toxic chemicals. Candida in the gut is like having a leaky toxic chemical plant inside.   Your body does not need an internal leaky toxic chemical plant. Candida is in no sense "normal". The only reason we have Candida in such large quantities is because of all the antibiotics people take and because we eat antibacterial chemicals as I explained in Chapter 2.

Candida breaks down and invades your intestinal lining

Candida is harmful in other ways. As we saw in a previous chapter, Candida makes toxic chemicals. This is only the beginning of Candida's destructive power. Another way in which Candida harms you is by synthesizing enzymes (protein tools) which break down the body's tissues. Scientific studies show that Candida invades body tissues after antibiotics are given.[7] These enzymes enable Candida not only to attach to your intestinal lining, but to invade that lining. We know that when a cancer patient undergoes chemotherapy, the membranes of the intestine are broken down. Yeast can and often does penetrate and invade the body. This infection is often lethal. Candida can and does kill many cancer patients each year.[8]

How the immune system responds to Candida albicans

So far, I have discussed how our immune system works, how part of our defense system is commensal bacteria, how antibiotics kill commensal bacteria as well as bad bacteria. I have discussed how killing the commensal bacteria gives space to the yeast, which attach themselves to our intestines. From there, Candida poisons your body by making toxic chemicals and invades by breaking down your intestinal lining.

Now we will explore how your immune system reacts to Candida. By now you must be asking, if our immune system is our body's defense department, how could it let Candida get so far?

The immune system treats Candida like a foreign invader and responds with both the offensive and defensive parts of the immune system. The immune cells generate the signals for inflammation. With the inflammation come offensive immune cells to fight and try to kill Candida. The swelling found during inflammation forms a barrier. The inflammation with its pain is a necessary part of fighting Candida. However, for reasons I will explore below, these responses are not very effective because Candida fights back against our immune system. Candida is a well-adapted and clever enemy.

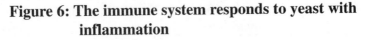

Figure 6: The immune system responds to yeast with inflammation

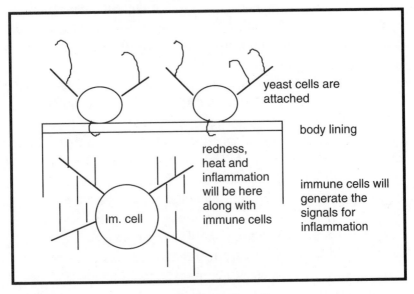

Before looking specifically at Candida, we must look at how immunologists learn the answers to how our immune system fights foreign invaders.

Immunologists cannot study how the immune system responds to Candida using people, because causing disease in people is unethical and prohibited by medical ethics and guidelines. So immunologists use the next best model, an experimental animal. They often use the mouse to look at how the immune system fights infectious microorganisms such as Candida. The mouse immune system is thought to resemble the human immune system closely enough to learn what happens when the human immune system fights Candida. In a major study, immunologists put Candida on the inner surface of the mouse mouth to see what would happen.

The scientists found that when Candida cells were placed onto a mouse's inner mouth surface, a vigorous inflammation in the mouth tissues occurred at the site of Candida placement.[9] This means that the mouse responded to the Candida both defensively (inflammation) and with immune cells which tried to kill the Candida. For a few days, the number of Candida organisms increased. Then an immune response developed more fully, which consisted of many immune system weapons. The number of Candida decreased to a small but measurable level. This meant that the immune system's offensive immune cells cleared the Candida mostly but not completely. Some Candida remained.

A month later, the researchers placed Candida cells on the inner surface of the mouths of the same mice who originally got the Candida.. An inflammation occurred again where the Candida was placed. This time the Candida cells were cleared more quickly. They came back to a small baseline level in a few days. Just as after the first time, the mouse immune system did not eradicate the Candida completely; some Candida remained. Again the immune system offensive weapons cleared most but not all of the Candida.

We learn from this experiment that the mouse immune system treats Candida as a foreign invader. The mouse immune system mounts a defense and an offense, but for some reason cannot defeat and clear the Candida completely. A baseline of Candida remains, which was not present before the experiment started.

Two important points to remember are these: first, by itself, the immune system cannot completely clear Candida. Second, this inability to eradicate a foreign invader is not the usual case when the immune system fights a foreign invader. Usually the foreign invader is destroyed completely. A third important finding is that not only does the Candida remain but it continues to provoke an immune response. Scientists found immune cells (unknown type) in the underlying layers of the tissues of the inner mouse mouth, even after the majority of the inflammation had resolved. Presumably these immune cells were defending against the Candida.

So Candida can fight back against the immune system, and continues to provoke an immune response.

Figure 7: Inflammation when yeast cells are present

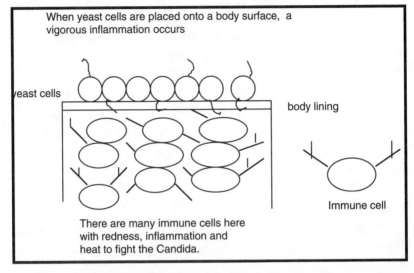

Figure 8: Yeast cells remain after inflammation resolves

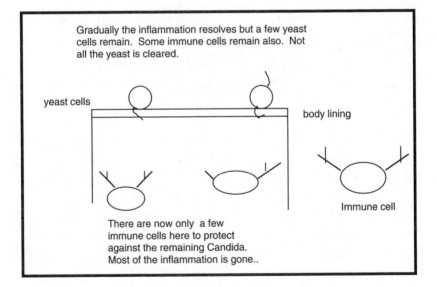

Why are these points important for us? Scientists believe that the immune system response of inflammation and other immune responses followed by partial eradication of Candida, as demonstrated in the mouse, is analogous to what occurs in human beings. Doctors see that Candida infections and increases in Candida level in people also lead to inflammation and immune system response. This kind of Candida increase will occur every time antibiotics are used. Antibiotics will remove the body's natural defense of good bacteria and allow Candida to flourish. Every time that Candida increases, there will be an immune response and inflammation.

Are human beings any better at clearing Candida than mice? In the natural state, Candida is not normally found in the mouth tissue of the mouse.[10] Smaller amounts of Candida coming in food, are not able to attach themselves in the mouse mouth. The mouse immune system is no weaker than the human immune system.

In this experiment, larger amounts of Candida were applied directly to the mouse mouth to learn how the mouse immune system handles such a challenge.

These studies on immune system response to Candida help us understand illnesses involving significant inflammation

As we saw above, inflammation is a major part of the immune response to foreign invaders. Inflammation is also a symptom of many hard to treat illnesses such as ulcerative colitis, psoriasis, and Crohn's disease. As we saw above, inflammation is also a major part of the defense against Candida. As long as Candida is present, immune cells will defend against it. These immune cells can generate the signals for inflammation. The studies involving placing Candida on a body surface of the mouse show that inflammation will be generated every time there is more Candida or more Candida activity.

What this means to humans is that any time Candida increases, such as when antibiotics are given, there may be a full blown immune response with inflammation wherever the

Candida is. This may be in the intestine. Such inflammation is extremely painful. This may be in the vagina; certainly the inflammation is painful and itchy. Such immune responses with inflammation would not be as severe if the immune system could clear Candida completely. Then whenever antibiotics were given, the Candida would be starting out from zero and the immune system would not have to work as hard to clear or try to clear the Candida. As we saw in the mice, however, the Candida is not cleared completely. Candida starts off being present when a second (or third and on) antibiotic is given. Immune cells are already there, ready to generate an immune response with inflammation. Yet the immune system still cannot clear the Candida.

Remember, too, that antibiotics are not the only problem. Recall that the food we eat also makes room for the yeast by killing bacteria, and that the food we eat can help yeast grow even better by providing growth factors for yeast. Even without antibiotics, Candida can grow because of the foods we eat which help it grow. Antibiotics of course clear out room for the yeast.

This cycle occurs because Candida fools the immune system. The immune response and its signals for inflammation will be present longer than if the immune system could promptly clear Candida.

In the mouse experiments, the inflammation mostly cleared. In human beings, wherever there is Candida, there is the potential for an immune response with inflammation. We still have not answered the question you must be asking; why does the body's immune system not clear the Candida completely?

Why your immune system cannot clear Candida: Candida fights back by tricking the immune system

The basic problem we can see is this; our immune system somehow cannot effectively clear Candida. So Candida remains in our intestinal tracts. If we eat the wrong foods, we feed it and help it grow.

But our immune system should do a better job. The reason our immune system does not work well against Candida is that Candida uses tricks to evade our immune system. Candida fights back.

This is not a new idea. Scientific studies show how Candida tricks our immune system. Several lines of research show that Candida has many tricks to evade the body's immune system. I will now discuss these studies.

To understand how Candida resists the immune system, we need to know that Candida has an outer covering called a capsule. This capsule is a complicated structure into which Candida has packed as many tricks as possible. Embedded within the capsule are many mechanisms which allow Candida to survive attack from the body's immune system. Although these are survival mechanisms for Candida, we can think of them as tricks to fool our immune system. We can think of this capsule as resembling a thick forest. The only way the body's immune cells can kill the Candida is to cut through part of this forest to strike at the Candida inside. But the forest is hard to cut away.

Using the forest analogy, the immune system would be foresters, people with training in cutting away trees. Their main tools would be saws. The immune system's tools are like saws to cut the trees and branches. However, the trees in the Candida capsule are not ordinary trees. The tree branches are covered with a gum which sticks to everything, making the saws useless. Besides watching out for the gum, the people doing the cutting have to hang onto their saws. Otherwise the branches of the trees in the capsule could grab the saws, take them away, and prevent them from cutting. Drawing the analogy further, the thick forest, the Candida capsule, can grab, bind and neutralize the immune cells' tools against the Candida. The capsule makes it much harder for the body's immune system to fight and kill Candida.

To continue the analogy, think of the tree branches on

Candida as having special structures, through which only certain types of saws will cut. Candida is more like a magical forest that can change the shape, size, and type of the trees and branches to fool the forester. Candida actually resists the immune system by changing the structure of the branches so that another saw is required. The immune cells bring one kind of saw and tell other immune cells to bring more saws, but by the time these cells and saws arrive, the yeast has changed structure and these saws will not work. In other words, Candida is a changeable target. Candida can change form, and can bind and neutralize immune system weapons.

If you had to walk through a forest looking for the right branch to cut and the branches kept changing and at any moment might bind your saw, you would have trouble. Your immune system has trouble with Candida. Candida can bind and neutralize immune system weapons.

The research supports this analogy.

The Science

Several lines of research show that Candida has many tricks to evade the body's immune system. Scientists did this research with whole animals and with immune cell cultures. These are plates of immune cells mixed with growth factors and kept under sterile conditions. In these experiments, immune cells were removed from the body. They were placed in plates or test tubes, then mixed with whole Candida cells or with the Candida capsule to see what happened. These experiments show that Candida's capsule can bind, or trap important immune molecules. By such binding Candida can interfere with immune cells. I will present the basic outlines of what we have learned from these experiments.

The yeast Candida interferes with the body's immune response to Candida in many different ways. One way to understand this interference is to look at what should happen between the body's immune cells and Candida. As I explained before, the immune system works using many cell types. The first main line of defense against foreign invaders is called phagocytosis. Phagocytosis is the scientific word for immune cells eating microorganisms. During phagocytosis, immune cells surround foreign invaders. These immune cells release enzymes to generate high energy molecules which destroy the foreign invaders.

The immune system is like an army with many different weapons and strategies to use against foreign invaders. This system is complex. For example in phagocytosis, immune cells must find the Candida, bind to it, and surround it. This process involves several different types of immune cells working together. Some of the immune cells should recognize Candida as a foreign invader, and make molecules, such as antibodies, which attach to foreign invaders. The antibodies attach to the yeast and other cells of the immune system use these antibodies to bind to, surround and phagocytize (or eat) the Candida. Another set of molecules, called complement, can also attach to foreign invaders. These molecules of complement are also used to assist phagocytosis.

Candida's thick capsule thwarts the immune cells

The first layer of the yeast Candida which the immune cells will contact is the capsule, a thick complicated structure. Important molecules of the immune system must bind to this capsule if they are going to begin to destroy the Candida. For example, prior to phagocytosis, a molecule (complement) is attached to Candida. This binding enables phagocytosis to take place. From research studies on mice, we know that mice which lack complement are hypersusceptible to Candida infections.[11] How does Candida thwart this process?

Candida interferes with the immune system and complement

Candida puts out on its surface a receptor for this complement molecule that is identical to a receptor found in normal immune cells. This receptor is specifically designed to bind one end of the complement molecule. The other end is for foreign invaders. Complement molecules bind to this yeast receptor. The Candida receptor can bind the complement in the same way the immune cell receptor binds complement. This is like the tree branch grabbing the handle of the saw, preventing the forester from cutting the branch. The problem is that when complement binds to Candida in this way, this molecule is literally pointed backwards. The end which sticks out is for foreign invaders, not for immune cells. Then it not only does not help, but actually interferes with the process of phagocytosis. The immune cell which is supposed to be using complement can only find the wrong end, because the part to which the immune cell is supposed to bind is bound to the Candida instead.[12] Thus Candida thwarts a main first line of defense, phagocytosis. In addition, scientists have shown that Candida makes much more of this receptor when there is more sugar around.[13] If Candida has more of this receptor, it is more resistant to immune system attack. Candida will make more of this receptor in people with diabetes, which may be one reason why diabetics have more yeast infections. To continue the analogy, more sugar helps Candida make more branches to grab the saw handles of the immune cells.

This receptor may do more than bind the complement. Complement circulates in the blood in an inactive form. In the laboratory, scientists have measured complement after adding some yeast capsule to blood. They found that when they add the yeast capsule to blood, the complement was all consumed and not available to do its job.[14] In other words, the yeast capsule may activate the whole complement system such that it is activated and used up. Then the complement is not available to help fight Candida at all.

Other ways in which Candida interferes with the immune system

Candida may have other ways of interfering with this process of phagocytosis, because even after the immune cells surround the Candida, the immune cells have trouble killing it.[15] Candida can actually remain viable inside of the immune cells which have already taken it inside and are trying to kill it.[16]

One of the ways immune cells kill the microorganisms they have already surrounded is by generating high energy molecules that act like bullets to poke holes in the outer covering of the foreign invader. One of the enzymes which generates high energy molecules binds to the capsule of Candida. This enzyme can kill Candida. However, scientists trying to figure out why Candida is not killed suggest that the enzymes are not binding to a live Candida cell. Rather these scientists suggest that throughout its lifetime, Candida releases pieces of its capsule. The enzyme, then might bind to a floating piece of capsule rather than the capsule covering the Candida organism.   If the enzyme is bound to a piece of capsule only, this enzyme does nothing to attack the Candida.[17] In this same study, the scientists found that pieces of the capsule also interfered with phagocytosis. These scientists put pieces of capsule in a solution with whole Candida cells and immune cells. When the pieces of capsule were present, less Candida was killed.[18] Candida's capsule is known to circulate at least in more severe Candida infections.[19] This effect of the capsule on phagocytosis could be very important at local sites of infection besides being important in serious systemic Candida infections.

The moving target defense

We saw from the previous discussion how Candida thwarts the immune system's first line of defense, phagocytosis. Candida puts out receptors to bind complement. Candida also thwarts the enzymes that produce "bullets" to pierce its capsule. Now we will see how Candida thwarts another line of immune attack by becoming a constantly changing target, so the body's immune cells cannot figure out where or what it is.

The next step that the body's immune system takes to destroy a foreign invader is to recognize the outside receptors of a foreign invader. Then a population of immune cells multiplies which can then recognize and kill that foreign invader. These immune cells call in bigger cells which can again eat and destroy foreign invaders but only if the right signals are present.

Candida has a number of defenses against these immune cells. The first trick is to change its outside. This is like the magical forest, where the trees are constantly changing. Normally, the immune system recognizes the outside receptors of the invading organism. Then signals are generated, and the cells which recognize those receptors multiply. These receptors mark the organism as one to be destroyed. Candida albicans can change the receptors which it is displaying, making it difficult for the body's immune cells to react appropriately. In essence, Candida albicans is a moving target, which changes its form. [20]

Interference with communication between immune cells

The interactions which go on between immune cells to result in multiplication of immune cells, a necessary step in responding to a foreign invader, are complicated. A number of types of cells must interact properly. What could interfere with this interaction? Besides being a moving target, Candida jams the immune cells' communication.

Immune cells also put out receptors, branch-like structures consisting mainly of sugar molecules. Immune cells literally make millions of these structures (during immune system development). These structures have individual shapes, some of which, like lock and key, connect with structures of foreign invaders. When this happens the immune cell is activated. Immune cells use similar branch-like structures to communicate with each other. These are made of sugars, but differ in certain ways from receptors which attach to foreign bodies. When other free floating sugar molecules (especially in chains) are around, these immune responses and the communication between immune cells can be interfered with. It is as if the sugar molecules get in the way.[21]

How is this information related to Candida? Let us look at what happens when Candida's capsule is degraded, or broken down. Remember that Candida's capsule is composed of complicated branching tree like structures consisting mainly of sugars, with a little bit of protein, collectively called mannan. In serious systemic Candida infections, portions of this capsule circulate in the blood.[22] One can find in the urine of pregnant women structures identical to very small portions of the Candida capsule. [23] Some researchers have suggested based on this type of information that Candida has the machinery to degrade its capsule and release pieces of it.[24]

When the Candida capsule is presented to immune cells (which are called macrophages), the capsule is degraded down to pieces consisting of 2 to 5 sugar molecules in a chain. In an experiment, these portions of capsule were mixed with immune cells. When these immune cells were given signals to grow, they did not proliferate (increase in numbers) properly.[25] In other words, the chains of a small number of sugar molecules from the Candida capsule suppressed the immune cells from growing. The general term is immunosuppression (more information is given in footnote 25).

In other experiments when chains of two or more sugar molecules are placed together with immune cells, there is significant suppression of the immune cells.[26] Suppression here means interference with immune cell communication and multiplication. As noted above, immune cells degrade the Candida capsule down to chains of two, three, four or five or more sugar molecules. Candida may also release such chains. We know from experiments that such sugar molecule chains suppress the immune system by interfering with communication. Therefore, it is highly likely that the chains of sugar molecules from the Candida capsule interfere with immune cell communication, leading to suppression of the immune system.

There is evidence to support this hypothesis. Such chains of sugar molecules have been found to circulate in women during pregnancy.[27] Such fragments interfere with immune cells.[28] We would expect such fragments to interfere with the immune response to Candida in human beings. Let me explain where we get this information.

This information comes from research on the immune system during pregnancy. One finding was that the urine of pregnant women contained factors which inhibit the immune system.[29] At first scientists thought that these factors were hormones excreted in the urine. But when these hormones were purified, the suppressive factors were found to be something other than the hormones.[30] Scientists did not know what the factors were, so they called the immune suppressants "contaminants." Later these factors were found to be identical with structures found in yeast capsules.[31] These structures are combinations of sugar molecules. However, a few of the same structures occur to some extent in the human body. The researchers did not say that the inhibitory factors in the pregnancy urine were from yeast. These structures could have been at least partially from the human body's own turnover of structures (special proteins). For the record, I note that for the researchers' studies, they derived their experimental chains of sugar molecules from yeast, most likely because it was the most convenient.

Nonetheless, these chains of sugar molecules from the yeast capsules inhibit the body's immune system. Either these pieces of yeast capsule or something identical to them are found in the urine of pregnant women.

Is the yeast capsule circulating?

Pieces of yeast capsule circulate in people with severe Candida infections. We do not know for sure how much yeast capsule is released in people who have less severe Candida infections. However, the body has no incentive to circulate something which inhibits the immune system. The yeast has a tremendous incentive to release pieces of its capsule because then it can inhibit the immune system and have a home. The information about sugar molecules in the urine of pregnant women is important because most likely the chains of sugar molecules which are found in their urine are of yeast origin. These pieces of yeast capsule may circulate and inhibit the immune system in non-pregnant women and in men as well.

Part of how the immune system deals with foreign invading microorganisms is degrading them down to a form in which they are harmless and the parts can be excreted. When Candida's capsule is degraded, parts are released which suppress the immune system's ability to kill the Candida.

In other words, the capsule of Candida contains structures within it which, when they are degraded in the right way, can significantly interfere with immune cell function. This information may help us understand why Candida can grow inside the cells which are trying to eat them. As the immune cells degrade the Candida, pieces of the capsule are released which are immunosuppressive, or bind the very molecules which the immune cells are using to destroy the Candida.

Now we know that scientists have discovered at least three ways in which the yeast and the yeast capsule thwart the body's immune system. The yeast capsule binds complement. The yeast can change the receptors it displays and when either it degrades its capsule or the immune cells degrade its capsule chains of sugar molecules are released which interfere with immune cell communication. So we can begin to understand why the immune system has such a hard time fighting Candida.

There are most likely other ways Candida resists the immune system which await discovery. Let me mention one more way in which the yeast resist the body's immune system.

Candida can also activate immune suppressor cells

When pieces of the Candida capsule were injected into a mouse, a population of immune cells called suppressor cells (immune cells which downregulate or slow things down) was generated. In special experiments, these immune suppressor cells prevented other mice from mounting an immune response to Candida.[32]   Clinical observations in human beings show that the same can occur in people. Doctors have observed that in some patients with significant Candida, they find general immunosuppression or lack of immune responsiveness to Candida and other foreign invaders.[33]

In other studies, scientists have found pieces of the Candida capsule in the blood of people who have chronic Candida on body surfaces.[34]   When pieces of this capsule are isolated from the blood, and are placed in cultures of immune cells, the cultures of immune cells are suppressed and do not grow properly.[35]

In other words, someone who has much yeast and who then thinks that he or she cannot resist any infection, is right. The yeast hurts the body's ability to fight infections.

Collectively, these studies show that Candida has tricks to evade the body's immune system. Studies show that Candida uses its capsule to resist and suppress the immune cells. The extent to which such immunosuppression occurs in human beings depends on the patient. In people with observable Candida, lowered immune responses to Candida and other foreign invaders can be shown on skin tests.[36] These depressed responses indicate that immunosuppression to both Candida and other foreign invaders is taking place in people with significant Candida infections.

Whether there is an obvious Candida infection present or not, perhaps the best way to think about Candida and all of its tricks is to go back to the study on the Candida infection of the mouth of the mouse. The mouse can clear most of the Candida but not all. Candida's tricks enable it to stay as a presence ready to grow when conditions allow it to grow, especially after the use of antibiotics.

The immune system continues to fight back

The immune system is a persistent guardian of our internal welfare. Just because Candida fights back does not mean our immune system gives up. Our immune system still has to be on the alert against Candida. Candida has the power to invade and kill.

Unfortunately, the foods we eat can also help Candida suppress our immune system. Everyone who has studied high school biology knows that the body's digestive system breaks down whatever food comes in. Won't the body's digestive system break down the yeast capsule into chains of sugar molecules?

Isolated yeast capsule, both from Candida and from bread yeast, prepared in the right way, inhibits the ability of immune cells to grow.[37] This effect in the laboratory depends on exactly how the yeast capsule is prepared [38] but yeast capsule in the diet could inhibit the body's immune system.

The common food additive malt contains chains of two and three sugar molecules.[39] These chains are similar to fragments of degraded yeast capsule. When such fragments are placed with immune cells in research studies, the immune cells are suppressed.[40] These sugar chains from malt could also act to inhibit the interactions between immune cells. Dietary malt could be a source of immunosuppression.

We have also seen that Candida is very resistant and the offensive parts of the immune system cannot clear Candida completely. Scientific studies show how powerful yeast is at tricking our immune system. What happens as the immune system is continually confronted with a potentially lethal, invasive enemy that the immune system cannot kill completely? This is actually a very good question.

What can happen are major health problems. The kind of health problems which occur depend on which direction the immune system goes. If the immune system calls in other weapons designed primarily to fight other kinds of foreign invaders, allergy symptoms develop, as I discuss in Chapter 12.

If too much Candida is present, much inflammation will also be present. Many illnesses are characterized by too much, persistent inflammation. This is the problem in gut diseases such as Crohn's and ulcerative colitis. If we add on top that cells fighting Candida in the gut can circulate and start fighting Candida and generating signals for inflammation on the skin, then we have skin problems such as psoriasis, as will be discussed in Chapters 9 and 10.

The other option is for the immune system is to call in the reinforcements. These cells cause the whole immune system to work harder but less specifically. These cells are not so good at distinguishing body cells from foreign invaders. Sometimes body cells get attacked. This results in autoimmune disease, which I discuss in Chapter 11.

We now understand enough about your immune system to understand how the yeast Candida albicans can cause major health problems. You understand that our bodies constantly are besieged by foreign invaders, called microorganisms, including yeast and fungi, bacteria and viruses. Our immune system's job is to find these foreign invaders and destroy them.

Our immune system does this by phagocytosis (eating of the foreign invader), by killing virus infected cells, by helper cells making signals to generate antibodies or inflammation, and by producing antibodies and by generating inflammation. We also now know that part of our body's defense against foreign invaders is having good, or commensal, bacteria. These bacteria occupy the places that other foreign invaders, such as Candida, would occupy. When we take antibiotics, we kill the good bacteria, allowing yeast to adhere to our intestinal wall. The yeast gains a foothold. When we take more antibiotics, or when we eat the wrong foods, the yeast grow stronger.

Yeast release toxic chemicals which cause their own problems. I discuss these chemicals in another chapter. Yeast can penetrate the intestinal lining. In this section, however, I am concerned about problems yeast cause by interfering directly with the immune system.

We also have seen that when we put yeast in the mouths of mice, the mouse immune system attacks the yeast, but cannot completely clear the yeast. So each successive round of antibiotics, and each meal containing foods such as malt helps the yeast grow.

To find out why yeast is not cleared completely we looked at scientific studies about yeast. We found that yeast interferes with our immune system. Candida's capsule is like a thick forest. This forest gums up the immune system by binding important immune molecules. The trees change shape to fool the immune system. This is the moving target defense. Yeast may circulate decoy capsule pieces to use up the immune molecules. Yeast interferes with communication between immune cells.

But the immune system keeps fighting. Now let's find out what happens in our bodies when our immune system keeps fighting the yeast.

Notes

[1]Crislip, M.A. and Edwards, J.E., Jr. Candida albicans and Related Species. in *Infectious Diseases* ed. by S.L. Gorbach J.G. Bartlett and N.A. Blauklow. C. 283, W.B. Saunders, Philadelphia, 1992.

[2]Abu-Elteen, K. H. Incidence and distribution of Candida species isolated from human skin in Jordan. *Mycoses*, 42(4):311-7, 1999.

[3]Giuliano, M., Barza, M., Jacobus, N.V. and S.L. Gorbach.

Effect of broad spectrum parenteral antibiotics on composition of intestinal microflora of humans. *Antimicrobial Agents and Chemotherapy.* 31:202-206, 1987; Samonis, G., Anaissie, E.J. and G.P. Bodey. Effects of broad spectrum antimicrobial agents on yeast colonization of the gastrointestinal tracts of mice. *Antimicrobial Agents and Chemotherapy.* 34:2420-2422, 1990.

[4]Louria reviewed cases of serious complications resulting from giving antibiotics and having serious Candida develop. He gives many references to prior case reports.

Louria, D. B., Stiff, D. P., and B. Bennett. Disseminated Moniliasis in the Adult. *Medicine.* 41:307-329, 1962.

[5]Please see Physicians' Desk Reference from 1961 and from that era.

[6]Lehner (Lehner, et. al., 1972) studied patients with chronic mucocutaneous candidiases. This disease is characterized by persistent superficial Candida infection of the mouth, nails and skin, resistance to treatment, association with endocrine disorder in some and recurrent respiratory tract infections in others. In the skin is seen thickening with Candida infiltration and numerous inflammatory cells are present. None of these patients had greater immune responses on various tests even though the immune system was fighting the Candida. They had negative skin tests to Candida and some had in addition negative skin tests to PPD (tuberculin). Their immune cells did not make appropriate immune signals when stimulated (macrophage migration inhibition test). Lehner graded the severity of the immune defects in these patients.

Lehner, T., Wilton, J. M. A., and L. Ivanyi. Immunodeficiencies in Chronic Muco-cutaneous Candidiases. *Immunology,* 22:775-787, 1972.

[7]Tissue invasion also occurs in the gut. After antibiotic treatment, C. albicans can be found invading the intestinal mucosa (Kennedy and Volz, 1985). This invasion has also been demonstrated in an in vitro model. In a tissue preparation, C. albicans can invade vascular endothelium (Klotz, et. al., 1983); Kennedy, M.J. and P.A. Volz. Ecology of Candida albicans gut colonization: Inhibition of Candida Adhesion, Colonization, and Dissemination from the gastrointestinal tract by bacterial antagonism. *Infection and Immunity.* 49:654-663, 1985; Klotz, S.A., Drutz, D.J., Harrison,

J.L. and Huppert, M. Adherence and Penetration of Vascular Endothelium by Candida Yeasts. *Infection and Immunity.* 42:374-384, 1983.

8Odds, F. C. 1988. *Candida and Candidiases*, 2nd ed. Bailliere Tindall, London.

9Lacasse, M., Fortier, C. Chakir, J., Cote, L., and N. Deslauriers. Acquired resistance and persistence of *Candida albicans* following oral candidiasis in the mouse: a model of the carrier state in humans. *Oral Microbiology and Immunology.* 8:313-318, 1993.

10Lacasse, M., Fortier, C., Trudel, L., Collet, A., and N. Deslauriers. Experimental oral candidiases in the mouse: microbiological and histological aspects. *J. Oral Pathol. Med.* 19:136-141, 1990.

11Morelli, R., and L. T. Rosenberg. Role of complement during experimental Candida infection in mice. *Infection and Immunity.* 3:521-523, 1971.

12Gilmore and colleagues (Gilmore, et. al., 1988) showed that Candida albicans has the receptor for and binds a fragment of complement (iC3b). This fragment helps immune cells phagocytize (eat) microorganisms such as Candida. The receptor for this fragment is found on immune cells such as neutrophils. The fragment normally binds directly to microorganisms such as Candida and once it is bound, the neutrophil can more easily eat the Candida. However because Candida has the receptor for this fragment, the fragment binds in the wrong way (the fragment is literally backwards) and then the neutrophil is inhibited from phagocytizing the Candida. Gilmore and colleagues showed that when the yeast transform into their more virulent form, they have more of these receptors and then the immune cells cannot eat them as well. Also when there is more glucose (sugar) around the Candida makes more of these receptors. In a later paper at the same lab, Hostetter and colleagues (Hostetter, et. al., 1990) showed that growing Candida in more sugar leads to the Candida having more of these complement receptors and then the immune cells cannot eat them as well. These workers suggest that the presence of these receptors on Candida and the increase in their numbers when there is more sugar around is the reason diabetics are more subject to Candida infections. Gilmore, B. J., Retsinas, E. M., Lorenz, J. S., and M. K. Hostetter. An iC3b Receptor on

Candida albicans: Structure, Function and Correlates for Pathogenicity. *The Journal of Infectious Diseases*. 157(1), 38-46, 1988; Hostetter, M., K., Lorenz, J. S., Preu, L., and K. E. Kendrick. The iC3b Receptor on *Candida albicans*: Subcellular Localization and Modulation of Receptor Expression by Glucose. *The Journal of Infectious Diseases*. 161, 761-768, 1990.

[13] Gilmore, B. J., Retsinas, E. M., Lorenz, J. S., and M. K. Hostetter. An iC3b Receptor on *Candida albicans*: Structure, Function and Correlates for Pathogenicity. *The Journal of Infectious Diseases*. 157(1), 38-46, 1988; Hostetter, M., K., Lorenz, J. S., Preu, L., and K. E. Kendrick. The iC3b Receptor on *Candida albicans*: Subcellular Localization and Modulation of Receptor Expression by Glucose. *The Journal of Infectious Diseases*. 161, 761-768, 1990.

[14]Kind, L. S., Kaushal, P. K. and P. Drury. Fatal Anaphylaxis-Like Reaction Induced by Yeast Mannans in Nonsensitized Mice. *Infection and Immunity*. 5(2):180-182, 1972.

[15]Richardson and Smith (1981) isolated polymorphonuclear leukocytes, a type of white blood cell, from the blood. These are the cells which respond early to foreign invaders. They ingest bugs and generate and release high energy chemicals next to the foreign invaders which kills them. They mixed these leukocytes with Candida cells and asked how many Candida were still alive after being ingested for two hours. Approximately 75% of the Candida cells were still alive. When injected into other animals, these Candida cells killed them, even after being ingested by leukocytes for two hours.

In a second study, Cockayne and Odds (1984) followed a similar procedure and found that about 50% of yeast cells survived phagocytosis by PMN's. These authors point out that PMN's are not the only cells which attack yeast, so that their results should not be interpreted to say that Candida resists all immune cells which attack them. Nonetheless, Candida is resistant to one type of immune cell, the polymorphonuclear leukocyte, which is very important in fighting infections.

Cockayne, A. and F. C. Odds. Interactions of *Candida albicans* Yeast cells, Germ Tubes and Hyphae with Human Polymorphonuclear Leukocytes *in vitro*. *Journal of General Microbiology*. 130:465-71, 1984.

Richardson, M. D. and H. Smith. Resistance of Virulent and Attenuated Strains of *Candida albicans* to Intracellular Killing by Human and Mouse Phagocytes. *The Journal of Infectious Diseases.* 144(6):557-564, 1981.

[16]Cockayne, A. and F. C. Odds. Interactions of *Candida albicans* Yeast cells, Germ Tubes and Hyphae with Human Polymorphonuclear Leukocytes *in vitro.* *Journal of General Microbiology.* 130:465-71, 1984.

Richardson, M. D. and H. Smith. Resistance of Virulent and Attenuated Strains of *Candida albicans* to Intracellular Killing by Human and Mouse Phagocytes. *The Journal of Infectious Diseases.* 144(6):557-564, 1981.

[17]One of the main immune cells which attacks foreign invaders is called the neutrophil. The neutrophil "eats" foreign bugs such as yeast and bacteria. Neutrophils also release enzymes (tools made of protein) which synthesize the production of high energy molecules which can kill yeast and bacteria. Wright and co-workers (1983) found that this enzyme binds to the yeast capsule and when it is bound, this enzyme is more effective in killing yeast. The main problem is that if the yeast releases pieces of its capsule into the circulation, then the pieces of capsule bind this important enzyme and then this enzyme does nothing. Wright and colleagues suggest that this binding is a reason that Candida can evade immune cells such as the neutrophil.

Wright and colleagues (Wright, et. al., 1983) experimented with an enzyme released by neutrophils which can kill Candida. This enzyme, myeloperoxidase generates high energy molecules which kill Candida but for this enzyme to work best, it must attach to the yeast. The enzyme by itself was able to kill the Candida if it was bound to the Candida. If soluble yeast capsule was added to the mixture, then killing was less and if enough capsule was added, there was no killing of the Candida (because the enzyme was bound to the soluble capsule where it did nothing). When the immune cells were added, they killed the Candida about 50%. When the yeast capsule was also in the mixture, then the immune cells (neutrophils) could only kill about 6% of the Candida. The presence of pieces of capsule prevents killing of Candida. The concentration of soluble capsule in the blood is not as high as they used to prevent killing of Candida. However, at sites of local infection, the concentration of capsule could be high enough to prevent killing of Candida.

Wright, C. D. Bowie, J. U., Gray, G. R. and R. D. Nelson. Candidacidal activity of myeloperoxidase: mechanism of inhibitory influence of soluble cell wall mannan. *Infection and Immunity.* 42:76-80, 1983.

Wright, C. D. Bowie, J. U., Gray, G. R. and R. D. Nelson. Influence of Yeast Mannan on Release of Myeloperoxidase by Human Neutrophils: Determination of Structural Features of Mannan Required for Formation of Myeloperoxidase-Mannan-Neutrophil Complexes. *Infection and Immunity.* 43(2):467-471, 1984.

[18] Ibid.

[19]The Candida capsule can be found circulating in severe Candida infections. (Weiner and Young, 1976)

Weiner, M. H., and J. J. Yount. Mannan antigenemia in the diagnosis of invasive *Candida* infections. *J. Clin Invest.* 58:1045-1053, 1976.

[20] Ashman, R. R., Papdimitriou, J. M., Ott, A.K. and J. R. Warmington. Antigens and immune responses in *Candida albicans* infection. *Immunol Cell Biol.* 68:1-13, 1990

Brawner, D. L. and J. E. Cutler. Variability in expression of cell surface antigens of *Candida albicans* during morphogenesis. *Infection and Immunity.* 51:337-343, 1986.

Soll, D. R., Staebell, M., Langtimm, C., Pfaller, M., Hicks, J., and T. V. Gopala Rao. Multiple *Candida* strains in the course of a single systemic infection. *J. Clin. Microbiol.* 26: 1448-1459, 1988.

Slutsky, B., Stackell, M., Anderson, J., Risen, L., Pfaller, M. and D. R. Soll. White opaque transition: A second high-frequency switching system in *Candida albicans. J. Bacteriol.* 269: 189-197, 1987.

[21] Muchmore and colleagues investigated interactions between immune cells and foreign invaders in a series of studies. They noted that foreign cells put out receptors consisting of chains of sugar molecules and they found that our immune cells have

similar receptors on the outsides of such cells which also contain chains of sugar molecules. The immune cells use such receptors to know that a cell is foreign and should be destroyed.

In a series of experiments, Muchmore and colleagues examined whether sugar molecules could get in the way of these interactions between sugar receptors and they found that sugars could get in the way and inhibit the killing of foreign cells (Muchmore, et. al., 1980, 1981). They also experimented with chains of sugar molecules in a later experiment and they found that chains of sugar molecules (rather than individual sugar molecules) were much more inhibitory of the immune cells job of killing foreign cells.

Muchmore, A. V., Decker, J. M., and R. M. Blaese. Spontaneous Cytotoxicity by Human Peripheral Blood Monocytes: Inhibition by Monosaccharides and Oligosaccharides. *Immunobiology*. 158:191-206, 1981.

Muchmore, A. V., Decker, J. M., and R. M. Blaese. Evidence that specific oligosaccharides block early events necessary for the expression of antigen-specific proliferation by Human lymphocytes. *The Journal of Immunology*. 125(3):1306-1311, 1980.

In later work, they investigated the nature of this contaminant in the urine of pregnant women and combined this work with their work on the chains of sugars. They separated the urine of pregnant women into fractions and found which inhibited immune cells. They concentrated one fraction and found a molecule consisting of two sugar molecules bound together. They apparently did not have enough material from the urine for their experiments so they purified the same molecule from a different source, yeast cell walls so that they would have enough to test this double sugar molecule. They found that this molecule significantly inhibited T cells' responsiveness to all antigens tested from tetanus toxoid to streptokinase to Candida. This double sugar molecule also inhibited the interactions between T cells and monocytes. The double sugar inhibited much more than simple single sugar molecules. They noted that chains of such sugars were even more inhibitory than the double sugar.

They comment on the source of these inhibitory molecules in the urine of pregnant women, as that the double sugar may come as the breakdown product of larger molecules because there was a variety of sugar molecules in various chain sizes found in the pregnancy urine. They note that the double sugar molecule could be from a breakdown product of certain molecules in human beings. However, one of the reasons that the yeast cell wall is so hard to study is that it varies so much and the sugars in the yeast cell wall can be combined in chains of varying lengths. (Nelson, et al, 1991) The most likely source of the double sugar and other sugar chains would most likely be yeast molecules present in the intestinal tract. Nelson and colleagues (1991) suggest that the yeast itself sloughs off pieces of the capsule and the yeast may use tools (enzymes) to break down the capsule into smaller pieces which then may be released.

22The Candida capsule can be found circulating in severe Candida infections. (Weiner and Young, 1976)

Weiner, M. H., and J. J. Yount. Mannan antigenemia in the diagnosis of invasive *Candida* infections. *J. Clin Invest.* 58:1045-1053, 1976.

23 Muchmore and colleagues investigated interactions between immune cells and foreign invaders in a series of studies. They noted that foreign cells put out receptors consisting of chains of sugar molecules and they found that our immune cells have similar receptors on the outsides of such cells which also contain chains of sugar molecules. The immune cells use such receptors to know that a cell is foreign and should be destroyed.

In a series of experiments, Muchmore and colleagues examined whether sugar molecules could get in the way of these interactions between sugar receptors and they found that sugars could get in the way and inhibit the killing of foreign cells (Muchmore, et. al., 1980, 1981). They also experimented with chains of sugar molecules in a later experiment and they found that chains of sugar molecules (rather than individual sugar molecules) were much more inhibitory of the immune cells job of killing foreign cells.

Muchmore and colleagues investigated interactions between immune cells and foreign invaders in a series of studies. They noted that foreign cells put out receptors consisting of chains of sugar molecules and they found that our immune cells have similar receptors on the outsides of such cells which also contain chains of sugar molecules. The immune cells use such receptors to know that a cell is foreign and should be destroyed.

In a series of experiments, Muchmore and colleagues examined whether sugar molecules could get in the way of these interactions between sugar receptors and they found that sugars could get in the way and inhibit the killing of foreign cells (Muchmore, et. al., 1980, 1981). They also experimented with chains of sugar molecules in a later experiment and they found that chains of sugar molecules (rather than individual sugar molecules) were much more inhibitory of the immune cells job of killing foreign cells. Here they drew on earlier work in which they noted that the urine of pregnant women contains factors which inhibit the immune cells. At first, these compounds were thought to be hormones but Muchmore, A. V., and R. M. Blaese (1977) found that the inhibiting compound was actually a contaminant.

Muchmore, A. V., Decker, J. M., Blaese, R. M., and B. Nilsson. Purification and characterization of a mannose-containing disaccharide obtained from human pregnancy urine: A new immunoregulatory Saccharide. *Journal of Experimental Medicine.* 160:1672-1685, 1984.

[24]Nelson, R. D., Shibata, N., Podzorski, R. P. and M. J. Herron. Candida Mannan: Chemistry, Suppression of Cell-Mediated Immunity and Possible Mechanisms of Action. *Clinical Microbiology Reviews.* 4:1-19, 1991.

[25]Podzorski and his co-workers (1989) did a study of the effects of the Candida cell wall on immune cells. They isolated the cell wall by two different techniques and then degraded it. They then tested both the whole cell wall and the broken down cell wall for their effects on the immune cells. When the intact cell wall is presented to immune cells they multiply. This is the normal response of such immune cells to a foreign invader. However, when the cell wall is degraded into small pieces, the effect of these pieces on the immune cells is quite different. In this case when the pieces of the yeast cell walls are presented to the immune cells, the immune cells are inhibited from multiplying. Something in the pieces of the yeast cell walls prevents them from multiplying.

These authors went further and tested what happens when the yeast cell walls are picked up by these same immune cells for degradation. Then similar fragments of cell wall are released as degradation products from the immune cells. The writers note that the presence of these yeast wall fragments constitutes a mechanism for yeast immunosuppression.

Podzorski, R. P. Herron, M. J., Fast, D. J. and R. D. Nelson. Pathogenesis of Candidiasis: Immunosuppression by Cell Wall Mannan Catabolites. *Archives of Surgery.* 124:1290-94, 1989.

[26] Muchmore and colleagues investigated interactions between immune cells and foreign invaders in a series of studies. They noted that foreign cells put out receptors consisting of chains of sugar molecules and they found that our immune cells have similar receptors on the outsides of such cells which also contain chains of sugar molecules. The immune cells use such receptors to know that a cell is foreign and should be destroyed.

In a series of experiments, Muchmore and colleagues examined whether sugar molecules could get in the way of these interactions between sugar receptors and they found that sugars could get in the way and inhibit the killing of foreign cells (Muchmore, et. al., 1980, 1981). They also experimented with chains of sugar molecules in a later experiment and they found that chains of sugar molecules (rather than individual sugar molecules) were much more inhibitory of the immune cells job of killing foreign cells.

Muchmore, A. V., Decker, J. M., and R. M. Blaese. Spontaneous Cytotoxicity by Human Peripheral Blood Monocytes: Inhibition by Monosaccharides and Oligosaccharides. *Immunobiology.* 158:191-206, 1981.

Muchmore, A. V., Decker, J. M., and R. M. Blaese. Evidence that specific oligosaccharides block early events necessary for the expression of antigen-specific proliferation by Human lymphocytes. *The Journal of Immunology.* 125(3):1306-1311, 1980.

[27] Muchmore and colleagues found that the urine of pregnant women contain factors which inhibit the immune cells. At first, these compounds were thought to be hormones but Muchmore and Blaese (1977) found that the inhibiting compound was actually a contaminant. This contaminant turned out to be chains of sugar molecules (see next footnote).

Muchmore, A.V. and R. M. Blaese. Immunregulatory
properties of fractions from human pregnancy urine: evidence
that human chorionic gonadotropin is not responsible. *Journal
of Immunology*. 118:881, 1977.

[28]In later work, Muchmore and colleagues investigated the
nature of this contaminant and combined this work with their
work on the chains of sugars. They separated the urine of
pregnant women into fractions and found which inhibited
immune cells. They concentrated on one fraction and found a
molecule consisting of two sugar molecules bound together.
They apparently did not have enough material from the urine
for their experiments so they purified the same molecule from a
different source, yeast cell walls so that they would have
enough to test this double sugar molecule. They found that this
molecule significantly inhibited T cell responsiveness to all
antigens tested from tetanus toxoid to streptokinase to Candida.
This double sugar molecule also inhibited the interactions
between T cells and monocytes. The double sugar inhibited
much more than simple single sugar molecules. They noted
that chains of such sugars were even more inhibitory than the
double sugar.

They comment on the source of these inhibitory molecules in
the urine of pregnant women, as that the double sugar may
come as the breakdown product of larger molecules because
there was a variety of sugar molecules in various chain sizes
found in the pregnancy urine. They note that the double sugar
molecule could be from a breakdown product of certain
molecules in human beings. However, one of the reasons that
the yeast cell wall is so hard to study is that it varies so much
and the sugars in the yeast cell wall can be combined in chains
of varying lengths. (Nelson, et al, 1991) The most likely
source of the double sugar and other sugar chains would most
likely be yeast molecules present in the intestinal tract.   Nelson
and colleagues (1991) suggest that the yeast itself sloughs off
pieces of the capsule and the yeast may use tools (enzymes) to
break down the capsule into smaller pieces which then may be
released.

Muchmore, A. V., Decker, J. M., Blaese, R. M., and B. Nilsson. Purification and characterization of a mannose-containing disaccharide obtained from human pregnancy urine: A new immunoregulatory Saccharide. *Journal of Experimental Medicine.* 160:1672-1685, 1984.

Nelson, R. D., Shibata, N., Podzorski, R. P. and M. J. Herron. Candida Mannan: Chemistry, Suppression of Cell-Mediated Immunity and Possible Mechanisms of Action. *Clinical Microbiology Reviews.* 4:1-19, 1991.

[29] Muchmore, A.V. and R. M. Blaese. Immunregulatory properties of fractions from human pregnancy urine: evidence that human chorionic gonadotropin is not responsible. *Journal of Immunology.* 118:881, 1977.

[30] Muchmore, A. V., Decker, J. M., Blaese, R. M., and B. Nilsson. Purification and characterization of a mannose-containing disaccharide obtained from human pregnancy urine: A new immunoregulatory Saccharide. *Journal of Experimental Medicine.* 160:1672-1685, 1984.

[31] See explanation in footnote 28.

Muchmore, A. V., Decker, J. M., Blaese, R. M., and B. Nilsson. Purification and characterization of a mannose-containing disaccharide obtained from human pregnancy urine: A new immunoregulatory Saccharide. *Journal of Experimental Medicine.* 160:1672-1685, 1984.

Nelson, R. D., Shibata, N., Podzorski, R. P. and M. J. Herron. Candida Mannan: Chemistry, Suppression of Cell-Mediated Immunity and Possible Mechanisms of Action. *Clinical Microbiology Reviews.* 4:1-19, 1991.

[32]People with significant Candida infections frequently have negative skin test responses to Candida and sometimes to other antigens. A negative skin test means that the immune system is weak. Under certain experimental conditions, pieces of the Candida capsule can be shown to induce suppression of the immune system.

Domer and colleagues (1989) and Garner and colleagues (1990) showed that if a mouse was injected with the yeast capsule (mannan) that a population of suppressor cells was generated which could then be transferred to another mouse. The second set of mice was immunized with Candida and then a Candida skin test was done. Normally then the mouse should respond with a positive vigorous skin test and normal mice do this. However, if certain immune cells are taken from the mice injected with the yeast capsule and given to the immunized mice receiving the skin test, these mice do not mount a good skin test response. In other words, the cells from the mice which received the yeast capsule, have suppressed the normal mice's skin test response.

A number of studies show suppression of the immune response to Candida, but these are the best studies because they were done in live animals.

Garner, R. E., Childress, A. M., Human, L. G., and J. E. Domer. Characterization of *Candida albicans* Mannan-Induced, Mannan-Specific Delayed Hypersensitivity Suppressor Cells. *Infection and Immunity.* 58(8):2613-2620, 1990.

Domer, J. E., Garner, R. E., and R. N. Befidi-Mengue. Mannan as an antigen in cell-mediated immunity (CMI) assays and as a modulator of mann-specific CMI. *Infection and Immunity.* 57:693-700, 1989.

33Lehner (Lehner, et. al., 1972) studied patients with chronic mucocutaneous candidiases. This disease is characterized by persistent superficial Candida infection of the mouth, nails and skin, resistance to treatment, association with endocrine disorder in some and recurrent respiratory tract infections in others. In the skin is seen thickening with Candida infiltration and numerous inflammatory cells are present. None of these patients had greater immune responses on various tests even though the immune system was fighting the Candida. They had negative skin tests to Candida and some had in addition negative skin tests to PPD (tuberculin). Their immune cells did not make appropriate immune signals when stimulated (macrophage migration inhibition test). Lehner graded the severity of the immune defects in these patients.

Lehner, T., Wilton, J. M. A., and L. Ivanyi. Immunodeficiencies in Chronic Muco-cutaneous Candidiases. *Immunology,* 22:775-787, 1972.

34Fischer, A., Ballet, J-J, and C. Griscelli. Specific Inhibition of In Vitro *Candida*-Induced Lymphocyte Proliferation by

Polysaccharidic Antigens Present in the Serum of Patients with Chronic Mucocutaneous Candidiasis. J. Clin. Invest. 62:1005-1013, 1978.

35 Ibid.

36Lehner, T., Wilton, J. M. A., and L. Ivanyi. Immunodeficiencies in Chronic Muco-cutaneous Candidiases. *Immunology*, 22:775-787, 1972.

37Nelson and colleagues (Nelson, et. al., 1984) investigated the effects of yeast mannan (from the yeast capsule) on the ability of immune cells to grow in response to stimulation. They found that yeast mannan from the common bread yeast inhibited the ability of immune cells to grow when they were presented with various stimuli. They thought that this effect may be due to copper contained in the yeast cell wall because when they removed the copper, the yeast cell wall no longer had such an effect. The structures of the yeast cell walls from the common bread yeast and from Candida are similar and Candida cell wall also has such effects.

Nelson, R. D., Herron, M. J., McCormack, R. T., and R. C. Gehrz. Two Mechanisms of Inhibition of Human Lymphocyte Proliferation by Soluble Yeast Mannan Polysaccharide. *Infection and Immunity,* 43(3):1041-46, 1984.

38 Ibid.

39Briggs, D. E., Hough, J. S., Stevens, R. and T. W. Young. Malting and Brewing Science, Volume 1 Malt and Sweet Wort, Chapman and Hall, New York, P.88-89, 1981.

40 Please see references in footnote 26.

Chapter 9

Crohn's Disease and Ulcerative Colitis: disorders involving chronic inflammation in the gut

---

In the previous chapter, I discussed how the immune system works. We saw that one response to Candida on a body surface or lining was inflammation and an immune response. In this chapter, I will discuss two major medical conditions characterized by chronic inflammation in the gut. These conditions are Crohn's disease, involving chronic inflammation in the small intestine, and ulcerative colitis, involving chronic inflammation of the large intestine. What happens when our immune system generates an inflammation, but the inflammation does not clear Candida completely?

Any increase in Candida such as may follow antibiotics or pregnancy, can lead to a full blown immune response and inflammation. The inflammation may mostly resolve, but what if the Candida cannot be cleared down to a low level? What if the Candida is very resistant and grows at a higher level? What if the person receives antibiotics continually or has several courses of antibiotics in a short period of time? Because the immune system cannot clear Candida, the inflammation persists. Inflammation is defensive. This defensive reaction becomes very important if the offensive immune cells cannot clear Candida. This persistent inflammation results in a number of major health problems.

Prolonged inflammation characterizes Crohn's disease and ulcerative colitis

In two debilitating medical conditions, the main symptom is prolonged inflammation. Patients suffering from ulcerative colitis have prolonged, extensive inflammation of the colon or large intestine. Patients suffering from Crohn's disease have prolonged, extensive inflammation of the small intestine. These disorders are painful and debilitating. People suffering from Crohn's disease may undergo surgery to remove parts of the small intestine that have become so inflamed that the intestine threatens to close off. However, the surgery does not resolve the Crohn's disease.

A common belief in the research community is that all this inflammation must result from a defect in the immune system. So the answer is to suppress the immune system by taking powerful immunosuppressive drugs. However, no one has ever identified any spontaneous defect in the immune system which leads to such illnesses. Another perspective is that the inflammation is a necessary body process, so there must be a cause. What could this cause be?

Based on the evidence we have about yeast, and what we know about the immune system, and how effective clinical treatment is, I propose that the inflammation is the sign of an immune system mounting an appropriate response to a foreign invader. The inflammation in the gut in either Crohn's or ulcerative colitis resembles inflammation in more short term infections. But the difference is that in both Crohn's disease and ulcerative colitis, the inflammation persists and is painful.

What causes the inflammation to not resolve? Why does the inflammation continue? Whatever is triggering the inflammation is still present. The body's immune system responds to this foreign invader with inflammation and offensive immune cells. Based on the research and my own clinical practice, I propose that a very persistent foreign invader triggers this immune response with inflammation. I propose that this persistent foreign invader is Candida.

What is my scientific basis? I described in detail in Chapter 8 how the immune system responds to Candida, and how Candida evades the immune system and does not respond well to the immune system's attacks. In brief, after antibiotics are given, Candida can adhere to the inside of the intestinal wall. The immune system will fight the Candida with inflammation and offensive immune cells. The intensity of the immune response--inflammation--is related to how long Candida has been in the intestine, and how much Candida is there. If a person already suffering from intestinal inflammation takes even more antibiotics, Candida will have even more room to grow. The immune system will again fight the Candida, generating painful inflammation.

Candida can resist the immune system, leaving inflammation as a defense

More than inflammation is required to clear Candida. I explained in Chapter 8 that inflammation is a barrier to prevent spread of infectious microorganisms. To kill Candida, the immune system must go after it with its more powerful offensive weapons. However, these are not effective against Candida because Candida has many tricks to resist and evade the body's immune system.

Candida's receptors bind important immune molecules, pointing them in the wrong direction. Candida can change form to fool the immune system. Pieces of the Candida capsule suppress communication between immune cells. These are all ways in which Candida resists the body's immune system's offensive weapons.

What is left then is inflammation. The inflammation persists because Candida resists and evades the other parts of the immune system, not due to some inherent defect in the immune system. As long as the research is focussed on the inflammation, it will never find an answer to what causes Crohn's or ulcerative colitis.

There are no experimental models of Candida which show that humans or animals clear the Candida completely. Candida is known to evade the immune cells. Inflammation may be one of the last defenses if other immune cells cannot clear Candida.

You and the research community could ask then why do only certain people get disorders such as ulcerative colitis? I do not know for sure. The most likely answer would be that people vary in their ability to clear Candida. People whose immune systems are better at clearing Candida will not likely develop disorders such as ulcerative colitis. People who cannot clear Candida are at much higher risk of developing such disorders.

Even for people who cannot clear Candida well, there is a good therapy available. The 4 Stages diet and anti-yeast medicine nystatin clear out intestinal yeast. Then the immune system does not have to fight Candida and the inflammation resolves. My proposal is not just theoretical. In my clinical practice, treating for yeast treats ulcerative colitis and Crohn's disease. The following case shows the effect on ulcerative colitis of treating intestinal Candida.

Case of ulcerative colitis

Betty

Betty, 65, complained that she had had colitis since the age of 5. She had started rectal bleeding at the age of 40. Before that she had frequent stools. She lost weight and the bleeding stopped the following year. The bleeding eventually came back. Eleven years prior

to the appointment she went to another doctor who put
her on prednisone (cortisone in a pill) and made her into
a "fat pig" (her words). She was on prednisone for
three and a half years. She went to another doctor who
got her off prednisone but then she developed asthma.
At the time of the appointment, she regularly had
diarrhea seven minutes after eating. Food went right
through her. She had rectal bleeding. She had
Cushing's disease, a disorder of having had too much
prednisone. She still had steroid bumps, a side effect
of too much prednisone, and she could not lose weight.
She stated that she was not eating excessively. She
craved meat because she was anemic. Pasta did not
disturb her much. Vegetables and fruit caused diarrhea.
She was short of breath and could barely walk across a
room. Walking even a short distance caused asthma.
She still required prednisone twice a year for asthma.
She had abdominal pain as if someone was hitting her
gut. The pain was sometimes sharp. Rocking helped
the pain. She also had vaginal bleeding and uterine
bleeding. She was having stools on average 5 times per
day, 10 times a day on a bad day and 2 to 3 times on a
good day.

She was taking hormonal treatments, an antidepressant,
and medications for high blood pressure, diarrhea and
colitis as well as using asthma inhalers.

She started a small amount of nystatin and the 4 Stages
diet. Even a small amount of nystatin caused diarrhea
so she stopped it, but she stayed with Stage 1 of the
diet. Within one week there was no more rectal
bleeding, and no more diarrhea. She now had two
formed stools per day. Her abdominal pain was gone
and her asthma was better. She was using her inhalers
every other day instead of twice a day. She was no
longer waking with choking at night and her breathing
was easier during the day. She had not had an asthma

attack even though she had had bronchitis once. Her mood was good and she had lost ten pounds. She was not snacking anymore and she had no desire for chocolate. This woman said that she was a happy camper and that she now had her life back again. She had no need to see me again because she was feeling good. The follow-up was over the phone.

Sixty years of colitis was gone in one week.

Case of Crohn's disease

Ann

Ann came to me at the age of 43 with a ten year history of Crohn's disease. She had had Strep throat and two root canals about six months previously. She had received antibiotics four times. The most recent course had ended three weeks ago. Then she developed back pain. She had had prednisone four times which had usually cleared the symptoms. Now her bowel movements were normal. However, a test of general body inflammation was elevated. She had gas pains and abdominal pain on eating, so she was not eating much. Patients with Crohn's disease have tremendous abdominal pain. Medications for abdominal pain were not helpful.

She had had antibiotics every winter for twenty five years and had never had difficulty. She had been five years on the birth control pill, off for one year, then back on for eight years. She was also a smoker. She did not mention headaches.

She started the 4 Stages diet and nystatin. She called a week later to say that her belly pain was much better but now she was having headaches. I tried to explain to her why this might be.

I call this problem going from chronic to acute. The reason for this result is that the body somehow adapts to Candida at the cost of reduced function and/or continual pain. A chronic problem develops. Then one no longer sees acute reactions such as headaches. When one starts to clear Candida, the chronic problem gets better. However, if one takes in foods which contain toxic yeast chemicals, one sees acute reactions such as headaches. The answer is to clear the diet of foods containing toxic yeast chemicals and to continue to treat the Candida. Then the acute reactions will go away. This woman was only partially following the diet. Even then her Crohn's still improved.

We can see from these two cases that when you clear out Candida from the gut by following the 4 Stages diet and taking nystatin, you can resolve the inflammation that characterize Crohn's disease and ulcerative Colitis.

Chapter 10

Skin Disorders and Yeast

Out of all the research on Candida related disorders, research has been most direct on Candida and psoriasis. Researchers have actually shown that Candida causes all the changes in the skin characteristic of psoriasis.[1] When Candida is injected into the skin of an experimental animal, the skin lesions of psoriasis, including scaling and thickening, develop. The authors of this study suggest that the scaling is a defense against the Candida.[2] For some complicated reasons of experimental design, the authors stop short of saying Candida is the cause of psoriasis.

You might legitimately ask why researchers have this major finding that Candida can cause psoriasis, yet ignore this result, causing psoriasis sufferers to use other more toxic treatments including coal tar, steroids and methotrexate. I do not know for sure why researchers would not wish to follow up their finding. One reason I would imagine is that the relationship between the Candida on the skin and the Candida in the gut is not generally appreciated. Cells which are fighting Candida in the gut can circulate and start fighting yeast wherever these cells find yeast. If one takes a narrow view and only looks at skin, one will miss this relationship.

Immune cells fighting yeast in the gut circulate and find yeast on the skin

In Chapter 8, we saw that the immune system cannot clear Candida. The immune system fights Candida and generates the signals for inflammation. Unfortunately, the inflammation persists because the immune system cannot clear Candida completely. Many people who suffer from skin disorders can identify with a state of persistent inflammation. They see such inflammation on their skin every day, in the form of eczema or psoriasis. How can Candida in the gut lead to chronic inflammation on the skin and skin disorders?

Unfortunately, as long as immune cells are fighting yeast in the gut, immune cells will be looking for yeast elsewhere in the body. Remember that immune cells circulate throughout the body. One place these immune cells might find yeast is on the skin. Is Candida found on normal skin? Candida may be found on normal skin, especially in skin folds.[3] Candida is much more likely to be found on skin after antibiotic use, after steroid use or in people with skin disorders.[4]

When the immune cells come into contact with Candida on the skin, these immune cells will generate the signals for inflammation. These immune cells will try to clear the Candida. But as we saw in Chapter 8, Candida has many tricks to evade the body's immune system. If the immune system cannot clear the Candida, inflammation will persist. If the inflammation occurs in the hair follicles, acne can result. If the inflammation is spread more generally, we can see eczema. When the inflammation becomes even worse, we see psoriasis. All of these problems result from the body's immune system fighting Candida on the skin. The immune system is unable to win so the inflammation persists.

Candida is also one of the main causes of persistent itching. Itching is a painful sensation that something must be removed from the skin and if there is no other way, one scratches. Your body would love to remove the Candida.

Cases of Psoriasis

James

James came to me when he was 30 years old. His main problem was psoriasis, worse on the legs. He had a family history of psoriasis. James had had psoriasis for nine years. It started after he had lost a job. James had used creams. His problem was better during the summer. His skin would clear with creams and exposure to the sun. James also suffered from intermittent heartburn. He had had penicillin for Strep throat several times.

James had started drinking alcohol when he was 14 and he had also used marijuana and cocaine. He had been sober for three years. Not drinking had helped his psoriasis. On physical exam I saw that James had psoriasis lesions on his legs with a few lesions on his arms.

James started nystatin and the 4 Stages diet. James returned a month later and said that he was feeling good. He found the 4 Stages diet boring, but had no problems with nystatin. His skin was clearing. He still had some itching. He thought that his heartburn was worse initially but was now better. On exam, the psoriasis on the backs of his legs was gone and the psoriasis on the front of his legs was clearing.

He came back two months later and said that he was feeling good and that his skin was continuing to clear. He had occasional itch. He had only occasional reflux (heartburn). When I looked at the front of his legs, his psoriasis lesions were indeed continuing to clear.

Tanyia

Tanyia came to see me at age 40. She told me she had a one year history of psoriasis which had come on two weeks after quitting smoking. She had psoriasis under both breasts and in her armpits, which had become infected with yeast. These areas were raw, and strong smelling with peeling skin. Cortisone creams cleared the condition temporarily, but not the rawness. Tanyia also had psoriasis lesions on her back, elbows and pubic area. Tanyia had had eczema as a child until the age of 30. She had been on oral contraceptive pills from the age of 12 to 19 for menstrual problems. She had regular periods now. She had taken antibiotics as a teenager for strep throat and ear infections and had taken tetracycline for a vaginal infection two years previously. She had been pregnant one time.

Tanyia started the 4 Stages diet and nystatin. I also prescribed topical nystatin cream to put directly on the problem areas. Tanyia came back three weeks later and said that use of nystatin cream on some of the psoriasis affected skin areas made them red and worse. However psoriasis areas on her elbows and legs were clearing. On exam her right elbow was much improved and her left elbow had partial clearing. The armpits were less red but the areas under the breasts were still raw. The back lesions were starting to clear at one edge. She was taking a half teaspoon of nystatin four times a day.

She came back six weeks after starting the treatment and said that her psoriasis was good, a "miracle." Her back had mostly cleared. Her scalp did not hurt. Under her breasts there was still itchiness and moisture and redness but now nystatin cream was helping these areas. Her comfort level was much better. Her skin was no longer splitting and bleeding. Her legs and pubic area were clearing. On exam the areas under her breasts were less red, more pink, with some cleared areas and there was less redness of her armpits.

She came back three months after starting treatment and the improvement had continued. Her psoriasis was mostly gone. A little remained on her lower back. There was still soreness and moisture under her breasts but the itchiness was gone. Now she was having headaches.

The next visit was a month later, four months after starting treatment. Tanyia reported that her skin was continuing to clear but her headaches were continuing. She was now having daily headaches. Her dose of nystatin was one teaspoon four times a day. She had a little eczema on her fingers. The skin under her breasts was much better. She stated that she was following the diet well. On exam, the skin under her breasts was clearing well with only a little redness present.

She called a month later to say that she stopped nystatin for two days and her psoriasis started to come back. She restarted the nystatin and the psoriasis began to resolve again.

I would comment that probably something in her diet was which was bothering Tanyia and causing headaches. She had probably unwittingly left something in her diet to which she was sensitive. I also find that the dose of nystatin cannot be above three eighths teaspoon, four times per day, without very good compliance with the diet. The more the yeast is cleared out, the more sensitive some people become to any yeast chemicals which come in.

Case of acne

Mario

Mario came to me at age 25 and told me he had acne breaking out on his neck. He had had acne as a teenager but had never broken out on his neck. He stated that he was now breaking out in blotches, never to this extent before. His treatment history was that he had taken tetracycline for five years up until four months previously. Tetracycline had controlled the acne at least for a while. Other problems were that drinking alcohol caused loose stools. He complained that he felt tired after eating. He had given up caffeine six months previously due to not having enough energy for workouts. His skin was broken out on his back and tetracycline had never helped this. He was taking acidophilus, several vitamins, digestive enzymes and golden seal root.

He started the 4 Stages diet and nystatin.

Mario saw me six weeks later and stated that his acne was better. His neck was better and his energy level was better. He was taking one half teaspoon of nystatin four times a day and had no problems with it. On exam his acne on his face, neck and back were all decreased. Since then his acne stayed much reduced.

A case of severe generalized itching

Joan

Joan, 54, came to me and told me she had generalized itchiness on her neck, arms and chest. This itchiness went back to her childhood and had worsened four years previously. She had had eczema as a child. At the time of the visit she also had itchy eyes and eyelids. She had scaling of her scalp at the back, and inflammation of the skin of her elbows, upper chest and at the waistline of her back. She had seborrhea (fatty secretions on the skin). She was using a tar shampoo to control scaling . She had low energy, but this had improved some. Joan had tried homeopathy for one year, but nothing had helped the itchy skin. Four years previously she had received a small amount of nystatin and some itching on her back had cleared.   In addition, Joan had hay fever and fibrous breasts. She was having hot flashes. She had used oral contraceptive pills in her late thirties and she had used antibiotics throughout her life. She was not taking any medications at the time and she had never been pregnant. Her gall bladder had been removed.

Joan started nystatin and the 4 Stages diet.

She came back three weeks later, taking one half teaspoon of nystatin three times a day. She complained of itching at the hairline of her neck and face. Her eyes were better. Her hot flashes had disappeared. She stated that her energy level was much improved. Her scalp was still flaking. I observed that the inflammation of the skin of her elbows was nearly resolved and her eyes were not swollen.

Joan returned three weeks after the second appointment. Her energy was still good. The itching on her body was better, but her neck and face were still bothersome. The scalp was less itchy. She could have sex more easily. Energy was still good. Her face appeared clear to me.

Joan came back in another three weeks, nine weeks after starting treatment. She was feeling much better. She still had a little itchiness in the morning, but it was mild. She stated that she felt 95% better overall. She noted that her itchiness would recur if she ate the wrong foods. Her hot flashes had come back, but they were not as often or as intense. In a later appointment she stated that the hot flashes would reoccur when the itchiness recurred.

I followed Joan's case for another five years. Joan found that her symptoms of itchiness would come back if she ate the wrong foods, such as vinegar or malt. The problem would disappear if she went back on the diet and continued to take nystatin. Otherwise she was free of her itching.

The case of Nell: A skin condition which baffled all the doctors

Nell

Nell came to me at the age of 67 to tell me of her skin problems. She had a scaly area of inflammation on her head about three inches in diameter and ringed red lesions on her back. The area on her head was bald and itchy. She had had many tests, including a test for lupus, and all was normal. Her doctors had suggested another dermatologist at a University medical center. A skin biopsy showed a lymphocytic infiltrate (immune cells were found in the area of inflammation). These rashes had started after Christmas (she was seen at the end of April). Her skin was itching and hurting. A topical steroid cream had helped. She was using Cetofil, a gentle medicated lotion.

Her other history and problems included that she had had zoster (shingles) 15 years before and had had major bowel surgery the past summer. She had taken antibiotics at that time. She had pain on eating. She currently had a sore throat and she had had a severe headache two days previously. She had a cold and she was still coughing a lot at night.

None of her local doctors knew what to make of her skin lesions. She did not even have a diagnosis. She also had a history of vaginal yeast infections. The physical exam showed the above skin lesions.

Nell went on the 4 Stages diet and nystatin. Nell lived out of state and called me frequently to tell me how she was doing. She came to see me four months later and told me that her back was almost clear and that her neck was clear. She was still using the Cetofil. She complained of blotches on her forehead starting two to three weeks previously and her eyelids were itchy. She was using hydrocortisone cream and Cetofil washes. She had some itchiness on her upper inner arms and in her groin. Tinactin, an antifungal cream, had helped briefly. She said her scalp was scaly again. She was taking a half teaspoon of nystatin four times a day. I suggested reducing the dose of nystatin to three eighths teaspoon four times a day.

Nell called me many times on the phone between appointments. I next saw her 11 months after the first appointment. She told me that her condition had finally improved. At that time she reported that the skin on her neck was clear. She had some redness on top of her head about the size of a quarter with no scales. She was still using Nizoral shampoo (an antifungal shampoo) and some Derma Smooth (flucinolone). She was still taking nystatin at three eighths teaspoon four times a day. By the following year the skin on her head was completely clear.

These cases truly are remarkable because patients who had suffered for months or years were able to clear up major skin disorders by treating for Candida. By modifying their diets and taking the anti-yeast medication nystatin, they conquered problems that standard medicine treats as incurable and chronic. Yes, these cases are remarkable, but are typical for skin problems. They are not isolated miracles.

In our own family, we have also seen this occur. My own son, who has been on the 4 Stages diet and nystatin for several years, developed severe eczema at one point. The main problem was that he was eating moldy plants outside. Other children with developmental problems have done such things. The dermatologist believed this case was so severe we got a call to enroll our son in a study. Instead, we tightened up our son's diet, increased the nystatin, and also gave another antifungal medication (ketaconazole) for two weeks, and the eczema disappeared. I would add that the dermatologist did not believe us and kept "assuring" us the eczema would return. Of course, it did not return in the subsequent 8 years except for a few patches when our son deviated from the diet.

Chronic skin disorders of psoriasis, eczema and itching are caused by a persistent inflammatory response against Candida on the skin. Clearing the yeast defeats these skin disorders.

Notes

[1]Sohnle and Kirkpatrick (1978) injected Candida into the skin of guinea pigs and found that the skin responded by increasing the rate of cellular proliferation. In other words more cells were generated at the base of the skin. The skin thickened within a few days, which is consistent with the immune system defense of delayed type hypersensitivity and then profuse

scaling developed. These authors noted that the scaling may be a defense mechanism by which the skin removes invading organisms such as Candida from the skin. In this particular study, both animals previously immunized to Candida and animals not immunized to Candida were used. The cells in the skin of immune animals proliferated more rapidly than the cells of skin of non-immune animals, indicating that increasing the number of cells is important in defense again the Candida. The authors stop short of saying that Candida infiltration is the cause of psoriasis because injection of compounds, such as ethanol, which cause inflammation lead to increased production of cells (but not scaling). According to these authors, Candida is not the only cause of increased cellular production in the skin. Nonetheless, these authors have shown that Candida can cause all the major changes of psoriasis.

Sohnle, P. G. and C. H. Kirkpatrick. Epidermal Proliferation in the defense against experimental cutaneous candidiasis. *The Journal of Investigative Dermatology.* 70:130-33, 1978.

[2] Ibid.

[3] Candida may be found on normal skin, especially in skin folds. Schaller, M., Schackert, C., Korting, H. C., Januschke, E., and B. Hube. Invasion of Candida albicans correlates with expression of secreted aspartic proteinases during experimental infection of human epidermis. *Journal of Investigative Dermatology.* 114(4):712-7, 2000.

[4]Candida albicans may be found on the skin in cases of dermatomycoses and this finding is more likely if antibiotics or steroids have been used or diabetes is present. Abu-Elteen, K. H. Incidence and distribution of Candida species isolated from human skin in Jordan. *Mycoses*, 42(4):311-7, 1999.

Chapter 11

Autoimmune Disorders and Candida

- *Multiple Sclerosis*
- *Numb Hands*
- *Rheumatoid Arthritis*
- *Fibromyalgia*

In Chapter 8, we looked at the basics of how the immune system works. We looked at how the immune system reacts to Candida. We saw that the immune system generates signals for inflammation as it attacks Candida. We also saw that the Candida resists the body's immune system. The immune system cannot clear the Candida completely. The immune system has to fight Candida continually.

In this chapter I will show you how understanding the immune system and Candida helps explain a group of illnesses called the "autoimmune disorders." Such disorders include rheumatoid arthritis, multiple sclerosis, fibromyalgia and other less common disorders such as scleroderma. In autoimmune disorders, the immune system attacks the body's own organs and tissues. For example, in rheumatoid arthritis, the immune system attacks the tissue of the joints. In multiple sclerosis, certain sites in the brain are attacked, leading to losses in nerve functions. At these sites of attack, doctors find serious inflammation.

Medical scientists have looked for many years for a defect in the immune system which causes autoimmune disorders. The scientists have not found such a defect.

Doctors prescribe drugs to relieve the symptoms of such disorders. Such drugs suppress and interfere with immune system function.

Does Candida play any role in these disorders? I believe so, based on treatment results. I have found that treating intestinal Candida reduces or even eliminates symptoms of these disorders. Why--- and how?

At first glance, the autoimmune disorders do not appear related to Candida. The immune system is attacking the body's own tissues. No infectious microorganisms or Candida are present at these sites of attack.

These disorders are complicated. To understand why Candida treatment leads to symptom reduction we need to bring together several pieces of information and lines of research.

At the sites of immune system attack, considerable inflammation is present, as if a foreign invader were present. Yet we can find no obvious foreign invader. The immune system is attacking the body's organs as if they were the foreign invader. What could trick the body's immune system into such attacks?

If we take the example of rheumatoid arthritis, the immune cells are attacking tissues in the body's joints. Normally very few immune cells are present in the joints. The immune cells have traveled to the joints. Something stimulated these immune cells to circulate from the blood or elsewhere into the joints.

Let's look at what is happening in the body. When foreign invaders enter, they circulate throughout the body. So the immune cells which fight these foreign invaders must also be able to circulate throughout the body to fight the foreign invaders, wherever they may be. Our bodies know that foreign invaders don't necessarily stand still. If a foreign invader is found in one place in the body, that foreign invader could spread. So the immune cells have to circulate. They must find the foreign invader wherever it is. This is the immune system's seek and destroy mission.

To further understand how our immune system works, we need to know a little about how our immune system recognizes foreign invaders. Doctors and medical laboratory workers can identify foreign invaders by looking at them under a microscope and by growing them in a laboratory. The body's immune cells do not have these kinds of tools. As we noted earlier, when discussing the immune system generally, the body's immune cells recognize foreign invaders by examining the structure of the foreign invader. Usually the foreign invader puts out different receptors, so the body's immune system knows that the foreign invader is foreign, and not a body cell.

The immune cells are constantly circulating looking for structures which are foreign and are not like the body's own cells. The immune cells are always making decisions that a foreign invader really is a foreign invader and not the body's own tissues.

When immune cells attack the body's own tissues, something is going wrong. The immune cells are circulating but they are stopping in the wrong place. In the rheumatoid arthritis patient, the immune cells are stopping in the joints and making the wrong decision, that the joints are a foreign invader. Then the immune cells start an attack leading to a painful inflammation. Why?

What if a foreign invader resembled the body's own cells? Then the circulating immune cells would have much more trouble distinguishing between foreign invaders and the body's own tissues.

What if a foreign invader resembled the body's own cells so closely that the body's immune system had trouble making this distinction? Then the immune system might attack the body's own organs.

These immune cells that attack the joints in a person with rheumatoid arthritis sense that the structures of the body's joints are exactly or almost exactly the same as a foreign invader. Once the immune cells decide that the structure of the joints are foreign, they do what they are supposed to do. They attack and cause inflammation.

As we saw in Chapter 8, the Candida cells resemble the body's own tissues very closely. As we have already seen, the body's immune cells are geared up to fight intestinal Candida. These cells circulate. If they find some structures which look almost exactly like Candida, they will attack, as if the Candida were present.

Does this really happen? The body's immune system is supposed to do a better job than this. To understand why the immune system does not do a better job, at least in some people with disorders such as rheumatoid arthritis, we need to remember that Candida is not static. Remember from Chapter 8 that Candida changes form constantly to trick our immune system. Candida also uses many other tricks to fool the immune system. So the immune system is fighting a sophisticated and constantly changing enemy that closely resembles our own body's tissues. Our immune system cannot just turn off its response to investigate further. Our immune system's job is to protect our bodies from foreign invaders, which can kill.

The immune system has to clear the foreign invaders. The only way to understand how the immune system can fail to make the distinction between foreign invaders and the body own organs is to understand that some foreign invaders are so powerful that the immune system has to put clearing the foreign invader as highest priority. The immune system sees attacking the foreign invader as much more important than occasionally not being able to make the distinction between the body's own organs and the foreign invader.

Candida is such a powerful foreign invader because it very closely resembles the body's own cells.

So the problem is that yeast cells resemble our own cells, making it possible for immune cells to confuse human cells with those of yeast. Let us develop these ideas further.

Let's talk about how yeast cells resemble human cells. Cells can look like each other if they put out the same receptors. Both yeast cells and human cells put out receptors, which are branch like structures which extend from the surface of the cell. Human cells communicate with each other this way. These receptors can receive hormones, such as thyroid hormone, which then tells the cell to do certain things. In the case of thyroid hormone, this might be "burn more energy and generate more heat." Some receptors, such as the connective tissue or laminin receptor anchor cells to connective tissue. Some receptors are for the immune system. Some receptors tell what kind of cell the cell is. All of the receptors are like unique fingerprints and tell the body that this cell is one of the body's own cells.

Foreign cells do not have these cell-identity receptors. Then the body knows that the cells are foreign. For example, when foreign tissue is transplanted into another body, the body recognizes the tissue as foreign; the right receptors are not there. The immune system then attacks the foreign tissues. Transplant patients take powerful drugs to suppress this immune response. The immune system is acting normally. It is supposed to recognize and attack foreign invaders.

Yeast cells fool the immune system because they display a number of human cellular receptors. Candida can display on its surface the human connective tissue receptor, called the laminin receptor, as well as other human receptors. The normal function of the laminin receptor is to allow human cells to anchor themselves to the body's membranes and to other cells. Candida uses this receptor in the same way, to anchor itself into the membranes of the human body, such as the inner intestinal lining. In the research world, displaying such receptors is known as molecular mimicry. Candida does this well and displays many human receptors.

Some of the other receptors Candida displays, in addition to the laminin receptor, are human receptors for immune molecules (complement receptors), and steroid hormone receptors .[1]

Displaying these receptors and most likely others also, makes Candida look to other human cells like a human cell. For example, both the laminin receptor and the steroid hormone receptor are found in the brain.[2] Displaying these receptors makes Candida look like a brain cell. The laminin receptor is widely distributed throughout the body. When Candida displays these same hormone receptors, connective tissue receptors, and immune receptors, Candida cells make themselves look like our own cells. They resemble other parts of the body, including connective tissue cells of the joints.

In other words, a major target of the body's immune system, Candida yeast, looks like our own brain cells, and the cells of our joints. If you think that this sounds like a set up for trouble, you are right.

But we cannot come to any conclusions about yeast and autoimmune disease yet. We need to know a few other things also. We need to know how the immune system handles chronic foreign invaders which have evolved to resemble body tissues and cells.

The immune system has difficulty recognizing Candida as foreign

As we saw previously, Candida is found in the intestinal tract. The immune system has trouble clearing it. As we saw in

Chapter 8, Candida has many tricks to evade the body's immune system. Nonetheless, the body's immune system knows that the yeast is not supposed to be in the intestinal tract. But recognizing the yeast as foreign is tricky because the yeast makes itself look like our own cells. Imagine if the American army had to pick out an invader as foreign if the foreign army was wearing American uniforms and displaying the American flag!

However, the development of the immune system takes account of the fact that some foreign invaders would make themselves look like our own cells. If the body's immune system could not handle a foreign invader which displays human receptors, these foreign invaders would have killed off the human race long ago. The immune system has developed to handle such foreign invaders.

The immune system produces many cells which are capable of recognizing many types of receptors and cells. The immune system produces all these different kinds of cells so that no matter what invaders attack the body, the body's immune system has cells which will recognize the invader and start the immune system on its search and destroy mission.

The developing immune system produces cells which can attack our own cells

The immune system produces many cells which can recognize our own cells, which I call our "self," and produce an immune response against them.[3] Some of these self-attacking cells are screened out during the immune system's development.[4] But many are not, leaving a group of immune cells which are potent and quite capable of attacking the body's own cells. Under normal circumstances, the body has ways of keeping such cells inactive.

When these self-attacking immune cells become activated, severe diseases result. When activated, they attack, resulting in severe diseases such as rheumatoid arthritis, multiple sclerosis, scleroderma, lupus and other autoimmune disorders.

What can activate such destructive cells?

Scientists have shown how such destructive cells become activated

One way to activate self-attacking immune cells is to present the body's own tissues to the immune system. Scientists do this by injecting such tissues into the body as if the injection were a vaccination.

If the body is immunized with its own proteins it will attack its own organs.[5] In other words, if the thyroid protein is presented by injection as a vaccination, the body's immune system and these destructive cells in particular will attack the thyroid gland. If a portion of the spinal cord is injected as a vaccination, the body will attack the brain and spinal cord resulting in a disease similar to multiple sclerosis.

These experimental injections are experiments, not normal living conditions. In real life, something prevents such destructive immune cells from becoming activated.

We also know that other cells of the immune system inhibit these self-attacking, destructive cells, so under normal conditions, these cells do not attack the body's own tissues. The cells that inhibit the self-attacking cells are inhibitory cells.

Scientists can remove these inhibitory cells by using fancy sorting techniques. When scientists remove the inhibitory cells, the experimental animal will attack its own tissues, resulting in autoimmune diseases very similar to those seen in human beings.[6]

In addition, when scientists remove the inhibitory immune cells, the rest of the immune system also attacks foreign invaders more vigorously.[7] In other words, these inhibitory cells not only keep the body from attacking itself, but also generally keeps the immune system a little less active when a foreign invader comes in.

So ironically these self-attacking, destructive immune cells help fight foreign invaders more vigorously, most likely especially foreign invaders which resemble the body's own cells. Let me explain in more detail.

Why do we have self-attacking, destructive cells?

Why are such destructive immune cells present? Autoimmune diseases cause terrible suffering. Why should our bodies possess the machinery for producing such illnesses?

The answer to this question provides insight into the cause of autoimmune disorders. Recall that the yeast Candida albicans can display human cell receptors and look like our own cells. Other microorganisms do this as well. If the body had no immune cells which could recognize our own receptors as part of a foreign invader, then microorganisms and yeast such as Candida could win every time. They would display human receptors, then invade and kill without any immune cells to stop them. But the Candida is unable to do this because the body's immune system has cells which recognize invading microorganisms which display our own receptors. To prevent these cells which recognize our own receptors as the enemy, from attacking all our own cells too, other cells inhibit them and keep them from attacking other self cells.

But we still do not know why our immune system attacks our own cells and what role Candida plays.

Candida acts like a vaccination

You already know that Candida is not a one time problem. In Chapter 8, I presented the studies done on mice. From those studies, we learned that our immune system does not clear

Candida completely. In addition, Candida keeps coming back. Every time a person takes antibiotics, or eats foods containing antibacterial chemicals, Candida grows. We saw in a previous chapter how the immune system continues to fight Candida. Could this continued growth and regrowth of Candida be anything like a vaccination?

The body's immune system is actually set up for brief periods of tissue destruction and presentation of the body's own receptors to the immune system. Every time there is an infection and/or trauma to the tissue, the body's own receptors may be presented to the immune system. But these brief presentations do not activate the very destructive cells to get an autoimmune illness started.[8] But let us look at Candida.

Recall now that Candida albicans displays our own receptors. Candida albicans is also locally invasive. Candida brings all of its receptors along as it invades.

In real life, Candida albicans puts out a number of destructive enzymes (tools) to invade our skin or tissue lining at specific places, which doctors call "locally," and this tissue invasion stimulates an immune response.[9] Scientists have demonstrated that this tissue invasion occurs in the mouth and in a number of experimental models.[10] This tissue invasion is found in the gut lining after antibiotic treatment.[11] Tissue invasion presents the Candida receptors to the immune system in much the same way as injection or vaccination would.

Each time antibiotics are used, Candida albicans will invade intestinal mucosa (linings) locally, stimulating an immune response. A series of antibiotics might be considered as a series of Candida vaccinations, presenting human cellular receptors to the immune system. After antibiotics, Candida albicans growth and local invasion will act like a vaccination, displaying the body's own receptors to the immune system. Our immune system reacts to Candida as it would to a vaccination.

Recall that if the body's own cells and receptors are presented as a vaccination for a long enough time, the body's destructive self-attacking immune cells will attack our own cells which display that receptor. This presentation of the body's own receptors will set the body's destructive cells in motion.

Candida may force the immune system to activate self-attacking cells

In addition, Candida albicans is a tough foe for the body. It may be that the body has to turn up the immune system to fight Candida and when that occurs, the destructive cells become more active.

Candida activates immune cells which can destroy our body's organs

The continued repetitive fight against Candida most likely activates the very cells which are capable of causing autoimmune disorders. Because Candida displays our own receptors, the cells of the immune system which are capable of fighting invaders displaying our own receptors are likely to be the cells which are called in to fight Candida.

If Candida were only present for very short periods of time and then only infrequently, Candida might not cause much activation of the destructive cells. But when people use antibiotics constantly, and when people eat the types of foods described in Chapter 2 which support yeast growth, Candida frequently is present. The use of antibiotics keeps on causing Candida to grow and re-present itself and our bodies' own receptors to the body's immune system. Each time the immune system requires time to fight this formidable foe.

Overall, allowing Candida to flourish on a body surface is like poking a hole in a body lining, sticking in structures which resemble the brain and says, "Come and get me." We should not wonder then that immune system defenders faced with this challenge sometimes attack the brain structures being poked through, in addition to the Candida cells attached to these brain structures.

The self-attacking destructive immune cells disappear when the Candida disappears

What describe above sounds like a hopeless, destructive, unstoppable downward spiral. In academic medicine it would appear that once the destructive cells are set in motion, nothing can stop them. This belief results in a lot of research to find ways to stop the immune system. Treatments to stop the immune system have side effects and do not necessarily work very well. However, these destructive cells can be turned off quite easily by simply removing Candida as an immune stimulus. I have found clinically that once Candida is treated, the autoimmune processes stop. To show this I will present some cases. These cases show what happens to patients with diseases such as multiple sclerosis when the Candida is treated and removed from the body.

Cases of multiple sclerosis

Kathy

Kathy, 43, had been diagnosed with multiple sclerosis (MS) about ten months prior to coming to my office. She had had an exacerbation of her multiple sclerosis after taking antibiotics (doxycycline) for dental work. Her MS symptoms included her right hand going numb, having "fireworks" in her right arm, having neck pain and trouble walking. She had had eye problems several years ago. Kathy also had sensations of her left leg feeling hot even if it was cold. She had trouble with stairs and bad fatigue. Difficulty with walking was the worst for her. She also had a vaginal yeast infection and migraines. Although her diagnosis of MS was relatively recent, Kathy believed her first episode may have gone back 16 years when she had restless legs after the birth of her second child. She had had a course of prednisone about six months prior to seeing me. Kathy reported the prednisone had helped for a few months. However, about a month before she saw me, Kathy took the antibiotic doxycycline. Her MS symptoms then worsened.

When Kathy was on oral contraceptive pills, she experienced massive mood swings. Kathy had taken many antibiotics as a child. She also had hiatal hernia for which she was taking two medications. She had three children.

Kathy started the 4 Stages diet and nystatin. She returned five weeks later. She reported that the aching in her legs was gone. She was having no problems in her right arm. Her neck was better, and walking was fine. Her heartburn was gone. Her migraines required aspirin, but no shots. The vaginal yeast infection was gone. Her neck hurt occasionally. She was still fatigued and had some mood swings.

Kathy was concerned about some of her remaining symptoms, but I pointed out to her that her multiple sclerosis symptoms were all gone except for occasional neck pain.

Kathy remained stable and pretty much symptom free for a year and a half.

Sadly, Kathy experienced a new multiple sclerosis attack at that time. Kathy was dragging her left foot and her left hand was weak. Kathy reported that she had not been observing the diet and had stopped nystatin. She had taken intravenous steroids which had helped briefly, but not significantly.

Unfortunately this can occur. A patient will be symptom free, then will relax the diet and stop nystatin. The yeast returns and so does the multiple sclerosis.

Tess

Tess, 45, complained of decreased appetite, sinus drainage, scratchy throat, shakiness if she did not eat, muscle tiredness, fatigue, light-headedness, and being bothered by certain sounds and fluorescent lights. Tess had been diagnosed with multiple sclerosis about twenty months previously when she had "inner ear problems". The room would spin when she turned her head to the right. MS was found on an MRI scan of the brain. Her multiple sclerosis symptoms included blurred vision, dizziness and arm tingliness. She did not stumble. She had had an abscessed tooth for which she took antibiotics about three months previously. Her inner ear problem had diminished before this time. Treatment of the abscessed tooth caused new MS symptoms for which she came to me. She had had a rash from penicillin. Tess had taken oral contraceptive pills for menstrual irregularities; the oral contraceptives had caused heart palpitations. She had a problem with vaginal yeast infections and she could not be around chemical smells. Tess reported that her belly did not "feel right."

Tess started the 4 Stages diet and nystatin. At three weeks, Tess was taking nystatin at three eighths of a teaspoon four times per day. She said that her muscles still fatigued easily but that this was better than before. She complained of a funny feeling in her head but no pain. Her appetite was fine. Her sinuses were better and she had no throat drainage. Her shakiness was better and the problems with sounds and lights were better. Her belly was fine. Her blurred vision and dizziness were better. She still had a vaginal yeast infection. Her energy level was better.

Tess felt good for the next year and half.

Later I found out that Tess ran out of nystatin about six months after the last appointment and had gone off the diet. A year after stopping treatment, Tess' light-headedness had returned. She had occasional numbness in her hands, and pain in her thighs. She had decreased energy.

Again, a patient who was virtually symptom free for a year and half decided to stop treatment. The yeast returned and so did the multiple sclerosis symptoms.

Doris

Doris came to me at the age of 47 and told me she had "mood swings." She had had multiple sclerosis symptoms for 26 years and had had a diagnosis for 24 years. She had been in a wheelchair for ten years. She also had cognitive problems, extreme fatigue, and vaginal infections. She had been on antibiotics all winter for a urinary tract infection and for ear infections. She cried easily and had hand numbness. She could stand but had balance problems. She had been gaining weight easily. She felt depressed and had been on Prozac, an antidepressant, for a long time. She had constipation and had problems with both bowel and bladder control. She had had many antibiotics as a child.

She started nystatin and the 4 Stages diet.

Doris came back one month later and said that she felt fabulous. Her depression had improved in one week.

Both her bladder and bowel control were better. Before, she said she could not void her bladder completely. She had no constipation. She said her hand numbness had improved. She could make decisions and a "fog" had been lifted. She had some joint pain. She had been up with a walker. Her accommodation to depth perception was quicker. She was still on Prozac, and she no longer craved fast foods.

Doris called me some time later to tell me that she had fallen, and for the first time in ten years she had been able to crawl. After that she made a decision that she could not tolerate the idea of being well again. She preferred life in her wheelchair and she stopped the treatment. I saw her later in a health food store, happy in her wheel chair.

All of these cases of patients suffering from multiple sclerosis show that within 4 to 6 weeks on the 4 Stages diet and nystatin, the patient can eliminate MS symptoms that sometimes have lasted for many years. As long as the patient remains vigilant about the diet and taking nystatin, they remain symptom free. If the treatment is stopped, symptoms can recur.

Case of Numb Hands

Janice

Janice, 35, had numbness in her arms and legs. The numbness varied in location, which is typical of multiple sclerosis. The numbness started about six weeks prior to her appointment with me. Janice had wet the bed during a dream the previous week and about eight months previously. She said that she was always tired. Her stress level had gradually increased and was difficult now. She had a history of "irritable bowel" for five months about eight years previously. During the five months of irritable bowel, she had severe abdominal pain. Janice had recently been on sulfa antibiotics which caused a yeast infection, which she said always happened when she took sulfa. She was taking penicillin for mitral valve prolapse, a condition of one of the heart valves. The penicillin also caused vaginal yeast infections. She was allergic to all perfumes and scents. She would develop diarrhea from drinking alcohol. She had diarrhea which alternated with constipation. Janice had used birth control pills in the past. Janice had taken antibiotics when she was younger. She had been pregnant two times.

These symptoms could have been the beginning of multiple sclerosis.

Janice started the 4 Stages diet and nystatin. She came back about six weeks later and reported that except for one episode of "labyrinthitis" or dizziness with bad fatigue about a week after seeing me, she was better now. She was following the diet and taking three eighths teaspoon of nystatin four times a day. The numbness in her arms and legs was totally gone. She had not wet the bed. All of her neurological tests had been negative.

Janice later told me that she felt that it had taken about six months for her symptoms to clear entirely, but everything had gone away.

Case of Rheumatoid arthritis

Mindy

Mindy, 53, told me that she had to stop playing tennis at the age of 40. She had been diagnosed at the age of 38 with rheumatoid arthritis. She had been tried on several medications and gold shots to which she developed an allergy. The pain in her knees, fingers and wrists was terrible, and her right elbow was deformed. In addition, her thyroid function was low. Her periods were erratic. She developed diarrhea from dairy products. She had vaginal yeast infections as a reaction to antibiotics. Mindy had taken antibiotics within the last year, which she did not tolerate. Mindy had also had an episode of pneumonia nine years previously for which she took antibiotics. She had three children. She was drinking two drinks per week. Mindy had had an episode of fatigue after some family stress a few years previously. She was on several medications for arthritis and for her thyroid. When I examined Mindy, I saw that her finger and wrist joints were enlarged.

Mindy started the first stage of the 4 Stages diet. She also began to take nystatin. Mindy came back six weeks later. She had stopped drinking alcohol, but was still craving chocolate. She reported feeling that she was doing better. She had had some bursts of energy. The swelling in her hands had gone down but had come back a little in the past week. Pain in elbows, knees and feet persisted. She said that her hands had improved the most. Her energy level was up. She was taking three eighths teaspoon of nystatin four times per day. She had not followed the diet strictly and still got some improvement.

A colleague of mine asked for help with rheumatoid arthritis which she had had for two years, starting at the age of 47. She was experiencing pain in her knees, elbows and hands. The problems were severe enough that she had enrolled in a special experimental study for which she had to travel out of state once a month. The study protocol did not help her much.

She started the 4 Stages diet and nystatin. She reported to me that the inflammation in her joints went down dramatically. She had less pain and increased movement in her hands. She also lost weight, which she had been trying unsuccessfully to do before.

Case of Fibromyalgia

Sherry

Sherry, 54, told me she had fibromyalgia for six years with a diagnosis four years previously. She had started with pain in her left eye. She sweated a lot. Her body had swelled all over, but swelling was now going down. Now she had pain in her arms, legs, feet and shoulders. She had occasional abdominal pain. She had stomach bloating. Her face was less swollen than it used to be. She would become discouraged and depressed. She had gained weight in the last three years and could not lose it. Sherry had done a little better with guaifenesin (an ingredient in cold medicines) and with massage therapy. She actually had come to see me to continue the guaifenesin which some people in California were recommending for fibromyalgia.

Sherry reported that her fibromyalgia problems had started after treatment with antibiotics for an infected tooth six years previously. She became fatigued, but previously she was very active. She had had bilateral carpal tunnel surgery two years previously. She had rib pain and was moderately obese. She had three children.

Sherry was both a supervisor and had to do manual work herself. She was finding the manual work difficult due to the pain.

Sherry started the 4 Stages diet and nystatin. A month later, Sherry told me that she had more energy. The extreme pain in her muscles was gone. Her muscles were still tender to the touch, but she could use her muscles without pain. She could walk on her heel with limited pain, but she could walk normally. She was not depressed and felt hopeful. She was no longer taking guaifenesin. She was taking three eighths teaspoon of nystatin four times a day. I observed that Sherry's mood was much brighter. She had only minimal swelling of her face.

Three months after the first visit, Sherry reported that overall she was continuing to feel better. She could use her legs much more easily. She had less pain. She could still feel some tenderness in her muscles but overall, this pain was less. She was able to do the manual work again that she needed to do. She also reported that the painful swellings in her muscles were much diminished in size, and Sherry had less overall body swelling. She was continuing to see a chiropractor and massage therapist. During this time, she had a single episode of breathing difficulties after eating chocolate. She had lost five pounds. She was taking three fourths teaspoon of nystatin four times a day.

Six years of pain and suffering had diminished in only three months. Sherry could now live the life she wanted to live.

Notes

[1]Calderone, R.A. and Braun, P.C. Adherence and Receptor Relationships of Candida albicans. *Microbiological Reviews.* 55:1-20, 1991. Bouchara, J-P, Tronchin, G., Annaix, V., Robert, R. and J-M Senet. Laminin Receptors on *Candida albicans* Germ Tubes. *Infection and Immunity.* 58:48-54, 1990.

[2]Laminin is a secreted protein which is a part of the basal lamina found in connective tissue. The laminin receptor is found in the brain of rats most notably on pyramidal cells and glial cells (Jucker, et. al., 1991). The steroid receptor can be found on oligodendrocytes and astrocytes, both types of cells found in the brain (Vielkund, et. al., 1990 and Jung-Testas, et. al., 1992).

Jucker, M., Kleinman, H.K., Hohmann, C.F., Ordy, J.M. and Ingram, D.K. Distinct immunoreactivity to 0kDa Laminin-binding protein in adult and lesioned rat forebrain. *Brain Research.* 555:305-312, 1991.

Jung-Testas, I., Renoir, M., Bugnard, H., Greene, G.L., and E. Baulieu. Demonstration of Steroid hormone receptor and steroid action in primary cultures of rat glial cells. *J. Steroid Biochem. Molec. Biol.* 41:621-631, 1992.

Vielkund, U., Walencewicz, A., Levine, J.M., and M.C. Bohn. Type II glucocorticoid receptors are expressed in oligodendrocytes and astrocytes. *Journal of Neuroscience Research.* 27:360-373, 1990.

[3]Wekerle, H., Bradl, M., Linington, C., Kaab, G. and K. Kojima. The shaping of the brain-specific T lymphocyte repertoire in the thymus. *Immunological Reviews.* 149:231-43, 1996.

[4]Ibid.

[5]Weigle, W.O. Analysis of auto immunity through experimental models of thyroiditis and allergic encephalomyelitis. *Adv. Immunol.* 30: 159, 1980.

[6]Sakaguchi, S., Sakaguchi, N. Asano, M., Ihoh, M. and M. Toda. Immunologic Self-Tolerance Maintained by Activated T Cells Expressing IL-2 Receptor alpha-Chains (CD 25). *The Journal of Immunology.* 155:11511164, 1995.

[7]Ibid.

[8]Weigle, W.O. Analysis of auto immunity through experimental models of thyroiditis and allergic encephalomyelitis. *Adv. Immunol.* 30: 159, 1980.

[9] Candida albicans produces a variety of pathogenicity factors, including adhesion molecules, proteinases, phospholipases, and lysophospholipases, which enable C. albicans to invade tissues locally (Ghannoum and Abu-Elteen, 1990). Once invasive, for example in the oral cavity, C. albicans will stimulate an immune response (Cawson and Rajasingham, 1972).

Cawson, R.A. and Rajasingham, K.C. Ultrastructural features of the invasive phase of Candida albicans. *Br. J. Derm.* 87:435-443, 1972.

Ghannoum, M.A. and Abu-Elteen, K.H. Pathogenicity determinants of Candida. *Mycoses.* 33(6):265-282, 1990.

[10]ibid.

[11]Tissue invasion also occurs in the gut. After antibiotic treatment, C. albicans can be found invading the intestinal mucosa (Kennedy and Volz, 1985). This invasion has also been demonstrated in an in vitro model. In a tissue preparation, C. albicans can invade vascular endothelium (Klotz, et. al., 1983).

Kennedy, M.J. and P.A. Volz. Ecology of Candida albicans gut colonization: Inhibition of Candida Adhesion, Colonization, and Dissemination from the gastrointestinal tract by bacterial antagonism. *Infection and Immunity.* 49:654-663, 1985.

Klotz, S.A., Drutz, D.J., Harrison, J.L. and Huppert, M. Adherence and Penetration of Vascular Endothelium by Candida Yeasts. *Infection and Immunity.* 42:374-384, 1983.

Chapter 12

Allergies and Yeast

People with intestinal Candida are more likely to have allergy symptoms or have worse allergy symptoms than if they did not have Candida. Why? As the immune system fights Candida, allergy symptoms can worsen.

First, what are allergy symptoms? For many people, the answer is obvious. People sneeze or have a runny nose around freshly mowed grass, or when a certain flower or weed is in bloom. Other people may have problems if they eat certain kinds of foods. Some people may experience rashes or even life threatening closing of the airway from eating certain foods. People can be allergic to many different things from molds in the air, to cat dander, to eggs.

Allergies and the immune system

The immune system generates the signals which lead to allergy symptoms. No one knows for sure why.

Allergy symptoms are reactions which occur when a person is exposed to something environmental, usually either from food or air. These reactions can vary as we noted. The reason that we need to look at allergy symptoms and yeast is that such symptoms are sometimes life threatening.

Intestinal Candida can make allergy symptoms worse

Let us start with respiratory allergy symptoms. Intestinal yeast can make these worse. Anyone who has had a runny nose around a blooming weed knows that his or her nose is generating extra fluid. The signals for this extra fluid and for other allergy symptoms come from the immune system. Why all the fluid?

We have already seen that the immune system can respond to a foreign invader with inflammation. Here the immune system seems to be telling the linings of the respiratory system to generate fluid to carry something away. This something of course is the pollen, grass or mold the person breathes in.

Allergy tests show that many things may stimulate the immune system to generate the signals for fluid release and other allergy symptoms. A person may have allergic reactions to plants, molds, grasses, dust or foods.

When something foreign enters the body, the immune system responds by making antibodies. There are five classes of antibodies, also called immunoglobulins, abbreviated Ig. Several of the types of antibodies simply stick to things and make it easier to clear out whatever has come in.

One class of antibody is called IgE. IgE can bind to foreign products, but IgE also can cause release of histamine. Histamine causes pain, swelling and fluid release. People with allergic symptoms make more IgE than necessary for ordinary daily living.

Mold spores are common in the air we breath. They might stimulate the production of antibodies so our bodies can clear them out. A small amount of fluid may wash some mold spores away, but common mold spores should not cause production of significant amounts of IgE.

This problem is even worse in people who have excess Candida.

People with Candida have more IgE

Has increased IgE, the antibody producing allergic responses, been found more frequently in people with more Candida? The answer is yes. This result was found in a study of people with asthma, rhinitis (runny nose), and eczema (skin inflammation).[1]   In another study, people who are carrying Candida have skin test reactions which are characteristic of people who have much IgE.[2]

Why should people with Candida have more IgE?

In Chapter 8, we learned that the immune system cannot clear Candida completely. As long as Candida is present, the immune system is fighting Candida. Candida is a formidable, evasive foe. Research shows that persistent foreign invaders such as Candida could stimulate the immune system into using other powerful weapons designed primarily to fight other kinds of invaders.

To fight Candida, the immune system has other weapons

The immune system is confronted repeatedly, with Candida, adhering to body surfaces. We know from Chapter 8 that the immune system is diverse and the immune system can generate several kinds of weapons to fight foreign invaders. Remember, the top priority for the immune system is to fight and clear the foreign invader. If the first weapons do not work, the immune system calls in other weapons, even if such weapons are not the most appropriate. We saw how Candida has many tricks to evade the body's immune system. What happens as the immune system fights this formidable Candida?

The first signal the immune system generates is for inflammation. For less challenging foreign invaders, inflammation and the offensive weapons which go along with inflammation clear out the foreign invaders. But Candida resists. What happens? Research shows that when there is a persistent foreign invader such as Candida, the immune cells can generate signals other than inflammation.[3] The more times the foreign invader invades, the more likely it is that such signals will be generated. One such signal is interleukin-4, IL-4, which then causes the production of IgE, the antibody of allergy. IL-4 and IgE are signals really for fighting a different enemy.[4] In nature, they are only called in because nothing else seems to be working.

We begin to see an answer to why the immune system is generating a strong reaction designed to fight a different enemy. The immune system is fighting Candida with whatever weapons it has. The antibody of allergy, IgE, is one such weapon. The longer the body's immune system has to fight Candida, the more IgE the body is likely to produce.

In other words, if Candida were present, the immune system would generate signals for inflammation, but as time goes on the immune system would also generate the signals for IgE production. IgE would cause a reaction to yeast that also includes fluid release and what we noted above as allergy type symptoms.

We all breath in yeast spores. People who have more Candida may react to breathing these spores. They would show both inflammation and fluid release. If this sounds like the beginning of asthma, I think you are right.

But other mold spores are not yeast. So why should the immune system react to them with the production of IgE? Other molds look like Candida and then cause release of IgE and allergy symptoms

The immune system recognizes foreign invaders by their structures. Candida has a capsule with a certain structure. Mold spores also have structures. The immune system identifies what is coming in by using immune cells which can detect the structures of what we contact.

The structure of Candida resembles the structures of other things such as common mold spores in the air .[5] Yeast spores are in the air,[6] and the structures of these spores and those of intestinal Candida are similar. Thus the structures seem similar to the body. The immune system reacts to both.

So when a fungus spore is inhaled, the body's immune system thinks that the spore looks like Candida and reacts as it has been reacting to Candida. Candida causes the immune system to use allergy type responses to it and to the things Candida resembles, such as common mold spores.

In other words, when the body's immune cells make IgE in response to Candida, the IgE can react against Candida but also against anything that looks like Candida. The problem is that Candida is structurally similar to other yeasts and fungus which are commonly found in the environment, in the air, and on the skin. Then these more common fungi will cause IgE to be produced and released, causing allergic reactions.

A group of scientists in Finland looked at the IgE antibodies to the yeast Pityrosporum ovale mannan (from the capsule) and the IgE to Candida. Pityrosporum ovale is commonly found on the skin. These researchers found significant cross reactivity in patients with atopic dermatitis (allergic inflammation of the skin).[7] Cross reactivity means that an antibody made to react against one microorganism reacts against a different microorganism. Because of this cross-reactivity, if a person has Candida and makes IgE (the antibody of allergy), the person will react allergically on the skin to the common skin yeast Pityrosporum ovale.

Other ways Candida causes allergic responses

Candida can act more directly as well. The capsule of Candida can cause release of histamine, one of the chemical signals causing allergy symptoms.[8] Injection of the Candida capsule in special experimental animals can cause an anaphylaxis like reaction.[9] Anaphylaxis is a fatal allergic reaction in which the windpipe closes off and the person cannot breath. Release of too much histamine results in anaphylaxis.

What about food allergies?

Food allergies are hard to understand. How could a person eat a common food such as egg and have his or her immune system react to that egg as if it was a foreign invader? Moreover, the immune system could react by producing IgE, which would cause an allergic reaction. Such an allergic reaction could range from a skin rash to a life threatening closing down of the airway. Why should the immune system react by releasing IgE and histamine in response to common foods?

The immune system would release IgE because the signals for such release are present. We have already seen that in some people, Candida can be such a signal. We also have seen that anything which structurally resembles Candida can be such a signal for IgE and histamine release.

Are these common foods similar in structure to Candida? Wheat structure resembles Candida.

One of the yeast proteins is similar in structure to one of the wheat proteins.[10] In one study, antibodies to wheat were noted to cross react with antibodies to yeast.[11] This means that to the immune system, both yeast and wheat have similar structures. If wheat protein comes into the body, the immune system may react as if yeast were in the body, provoking release of IgE and histamine. When such a reaction is severe, the histamine can close down the airways.

So when a common food such as wheat is present, the body's immune system can react as if Candida were present. The immune system then makes IgE and histamine, leading to an allergic reaction. A person who has significant amounts of Candida already has an immune system on hyper alert. That person reacts more strongly to a substance that resembles Candida than a person who does not have Candida. For some people, the substance might be wheat.

Unfortunately, there is not sufficient research to understand all of the foods to which people may react.

How does treating yeast help with allergies?

I have found that when we treat for intestinal Candida, allergy symptoms improve. Both respiratory and food allergy symptoms diminish. Let us look at some cases.

Cases

This case is presented because this little girl had asthma and wheezing, two symptoms of allergy.

Letisha

Letisha, age five, said that she did not feel well. Her mother reported that Letisha had had her first ear infection at six months and was then constantly on antibiotics. Letisha had had a white coated tongue one and a half years previously, indicating possible thrush, a yeast infection in the mouth, and had had a vaginal yeast infection up until a few weeks prior to the appointment. She had had two ear infections in the past three months, and had been treated with antibiotics. She had bad breath. She was low on iron, so she was taking iron. She had a chronic dry cough and had asthma diagnosed three months previously. She wheezed in the early morning. An asthma inhaler had helped briefly. She also had headaches and burning eyes, stomachaches, and occasional diarrhea. She craved sugar and candy. She was short tempered. She started fights constantly with her brother. Her mother had tried vitamins and acidophilus.

I started Letisha on the 4 Stages diet and nystatin. Letisha came back five weeks later. Her parents were happy with her progress. Letisha's sugar craving had diminished. Her cough had stopped. Her headaches had gone down. Her vaginal burning, abdominal pain and breathing were better. Her behavior was much better and she was getting along better with family members. She no longer wheezed. Her sugar craving was better. She still had some intermittent diarrhea.

Cheryl

Cheryl, 35, told me she had had allergies and chronic sinusitis. She had sinus congestion, itchy eyes and throat. Sinus congestion caused headaches. She also had a four year history of knee pain and sometimes hand pain. She had had fatigue for five years. Allergy shots had helped until one year previously when she had moved. An antibiotic would cause a vaginal yeast infection. She had taken antibiotics for bronchitis for 15 years. She stated that she had inhalant problems year round.

She started the 4 Stages diet and nystatin. She came back five weeks later. She stated that she was now having almost no headaches. She had more energy and more sex drive (not mentioned as a problem on the first visit). She stated that she still had intermittent sinus congestion but overall sinuses were better. Her knee pain was gone. She still had some postnasal drip. Her itchy eyes were a little better and her throat still itched some. Her allergy shots seemed to work better now.

We know from the research that people who suffer from Candida have worse allergy symptoms. Immune systems fighting Candida also attack everything that looks like Candida, which may include mold spores in the air, wheat, and other substances. Treating Candida dramatically reduces allergy symptoms.

Notes

[1]Savolainen, J., Koivikko, A., Kalimo, K., Nieminen, E., and M. Viander. IgE, IgA and IgG antibodies and delayed skin response towards Candida albicans antigens in atopics with and without saprophytic growth. *Clinical and Experimental Allergy*, 20:549-554, 1990.

[2]Koivikko, A., Kalimo, K., Nieminen, E., and M. Viander. Relationship of immediate and delayed hypersensitivity to nasopharyngeal and intestinal growth of *Candida albicans* in allergic subjects. *Allergy*. 43:201-5, 1988.

[3] Bradley, L. M., Duncan, D. D., Yoshimoto, K., and S. L. Swain. Memory Effectors: A Potent, IL-4 Secreting Helper T Cell Population That Develops in Vivo after Restimulation with Antigen. *The Journal of Immunology*. 150:3119-3130, 1993.

[4]The different enemy is parasites.

[5]Rantio-Lentimaki, A. Mould spores and yeasts in outdoor air. *Allergy*. 40:17-20, 1985; Pepys, J. Faux, J. A. Longbottom, J. L., McCarthy, D. S., and F. E. Hargreave. Candida albicans precipitins in respiratory disease in man. *The Journal of Allergy*. 41(6):305-318.

[6]Ibid.

[7]Lintu, P., Savolainen, J., Kalimo, K., Kortekangas-Savolainen, O., Nermes, M., and E. O. Terho. Cross-reacting IgE and IgG antibodies to *Pityrosporum ovale* mannan and other yeasts in atopic dermatitis. *Allergy*. 54:1067-1073.

[8]Nosal, R., Novotny, J., and D. Sikl. The effect of glycoprotein from *Candida albicans* on isolated Rat mast cells. *Toxicon*. 12:103-108, 1974.

[9]Kind, L. S., Kaushal, P. K. and P. Drury. Fatal Anaphylaxis-Like Reaction Induced by Yeast Mannans in Nonsensitized Mice. *Infection and Immunity*. 5(2):180-182, 1972.

[10]Camonis, J., et. al. Characterization, cloning, and sequence analysis of the CDC25 gene which controls the cyclic AMP level of Saccharomyces cerevisiae. *EMBO J.* 5:375-380, 1986.

[11]Vojdani, A. et. al. Immunological cross reactivity between Candida albicans and human tissue. *J. Clin Lab. Immunol.* 48:1-15, 1996.

Chapter 13

Recurrent Ear Infections and Recurrent Sinus Infections

Children who suffer from recurrent ear infections benefit from anti-yeast treatment. After treatment, they develop fewer and less severe ear infections. Adults who suffer from recurrent sinus infections also develop fewer and less severe sinus infections. To understand why, we need to know how Candida helps the microorganisms which cause ear and sinus infections. All of the following principles apply to adults with sinus infections as well as children with ear infections. To simplify the discussion, I am going to refer to children and ear infections here.

First, what is happening in these children who get recurrent ear infections?

Many young children seem to get one ear infection after another. Although many doctors and parents know that most ear infections are viral, doctors commonly prescribe antibiotics because bacteria cause at least some ear infections. In the typical cases that I see, the child takes a round of powerful antibiotics, only to have symptoms come back a few weeks later requiring yet another round of powerful antibiotics. Some children do not become well. Sometimes as more and more antibiotics are given, behavior deteriorates, ear infections continue and the physician has no answers. Why? What is helping the bad bacteria to come back?

Candida helps bad bugs to grow and come back

Bacteria are not alone at the back of the mouth and at the beginnings of the tubes leading up to the ears. There are many microorganisms. These include benign or helpful bacteria, viruses, and disease causing bacteria as well as the yeast Candida albicans. As we will see below, Candida interacts with and helps the bad microorganisms grow.

When a child takes antibiotics for an ear infection, the antibiotics kill both the benign and the disease causing bacteria, but not the yeast. The yeast remains. If a virus is causing the ear infection, the antibiotics kills the benign bacteria, but do nothing about the virus. The yeast also remains.

The remaining Candida then helps the bad bacteria grow back. Let me give some of the research on this point. Long before the days of complicated DNA experiments, researchers asked simpler questions. They asked whether bad microorganisms help each other and whether Candida helps bad bugs. The answer is yes. In the material presented in footnote 1, I cite studies in which researchers put Candida with a disease-causing microorganism. Together with the Candida, the disease-causing microorganism either grew better or survived longer.[1] My favorite example here is that when yeast and tuberculosis were grown together, they liked each other. The tuberculosis grew better with the yeast than without it. From this we know that the Candida capsule is a growth factor for tuberculosis. Candida also helps disease causing bacteria such as Staphylococcus, Pseudomonas, and Proteus.[2] Candida albicans helps Lactobacilli.[3] In other words, many studies have shown that the yeast Candida albicans helps other disease causing organisms grow better.

Although the main bacteria causing ear infections, Hemophilus, has not been studied specifically, yeast helps so many other disease causing bacteria that I suggest that if studied, Candida would help Hemophilus grow better. The same may be true for Candida and cold causing viruses.

So not only do antibiotics clear out the bacteria to make room for the yeast, but the yeast most likely promotes the growth of disease causing bacteria, leading to another ear or sinus infection.

What can be done about this yeast keeping the bad bugs growing where they can cause infection? To answer this question, we must first look at the relationship of yeast found in various sites of the body.

Most of the yeast in the body is found in the intestinal tract

Candida albicans is found at the back of the mouth and throat, in the intestinal tract and in the female genital tract. One finds the greatest amount of yeast in the intestinal tract. We can find yeast anywhere in the gut, which is very long with a lot of surface area. By comparison, less surface area is available to yeast in the throat, mouth, and female genital tract. The blood supply in the mouth and throat and in the female genital tract is high. Usually that means that infections are less likely. More blood means more immune cells can get to the area, making infections less likely. So why is there yeast at these sites?

Intestinal yeast helps yeast grow in the mouth

Yeast in the intestinal tract is quite active. As I discussed in Chapter 8, Candida evades the body's immune system. I will just summarize some of the ideas here. Candida has a capsule which contains several factors to help Candida resist the body's immune system. The evidence suggests that Candida releases tiny portions of its capsule into the circulation. These pieces of capsule interact with immune cells in such a way as to suppress these cells and prevent them from reacting vigorously to Candida albicans. Releasing tiny portions of yeast capsule prevents the body's immune system from clearing Candida from other body surfaces. In other words, when intestinal yeast releases pieces of Candida capsule, these pieces suppress immune cells which might attack the Candida in the mouth and throat.

Intestinal yeast must be cleared to reduce yeast in the mouth and throat

In summary, intestinal yeast enables yeast to grow in the back of the mouth and throat areas, and passages that connect with the ears, where the yeast can then help bad bacteria cause ear and sinus infections. Eliminating yeast from the intestinal tract is necessary to reduce the yeast in the throat and mouth.

The way to stop the process of recurrent ear infections is to clear the Candida out of the body, especially out of the intestinal tract.

Once you reduce the yeast burden in your body, you can break the cycle of ear infection, then antibiotic, then another ear infection. Children whose behavior has deteriorated after taking antibiotics will show better behavior. Adults can clear their recurrent sinus infections.

Cases

Terry

Terry, 20 months old, had had ear infections since the age of 5 months. He had taken the antibiotic Ceclor a month before the appointment. He had taken Augmentin a week before. He had taken many other antibiotics from five to 19 months of age.

The antibiotics were causing other problems. He had hives for three weeks after the last round of Ceclor. He was having other rashes which lasted from hours up to a day. He had recently had diarrhea with fever.

As an infant, Terry had been an easy baby. Now, his parents reported that his activity level had been "crazy" for the last few months. He was ripping things apart. For four months he had been up at night, screaming and yelling up to four times per night. He had a high tolerance for pain. He was understanding speech but his expressive speech was behind.

He had a rash on the day he was first seen. He was unhappy and was not smiling.

Based on my observations as a child psychiatrist, I viewed this child as at risk for developing a major developmental disorder.

Terry started the beginning of the 4 Stages diet and nystatin. Two weeks later, his mother reported that he was no longer so active. His mother could see a huge improvement in how he felt. He had no runny noses, no poking at the ears. He was sleeping better. He was still getting up at night but there was less screaming. In the office he was smiling and playing with toys.

Five weeks after the first appointment, Terry's speech had improved. He could say more words and he was putting two words together. His ears were free of infection.

Whatever risk there had been of major developmental disorder, was now substantially reduced. The cycle of ear infections and antibiotics was now broken.

The next case is a case of recurrent sinus infections. The same principles of yeast and ear infections apply to sinus infections.

Mark

Mark, 15, was now fatigued after being treated for recurrent sinus infections. At the time of the appointment, he was taking antibiotics for a sinus infection. He had had a sinus infections a few months previously. The antibiotic had not worked very well. The sinus infection had returned.

Mark had a history of many sinus infections. They had increased in frequency and severity about two years before the appointment. He was now having three sinus infections a year. During the sinus infections, he was "wiped out," with a small fever, pressure in his face, inability to focus, and fatigue. He also had allergy induced breathing problems, worse in both spring and fall.

Mark was having other problems too. Three months prior to the appointment he had developed stomach problems. His appetite was decreased. He was uncomfortable enough to leave school sometimes. He was burping, belching, and he had gas. These problems were severe enough to wake him at night. Then he might belch for hours. Overall, he said his stomach did not feel good. He had lost 10 to 12 pounds.

When Mark was 6 years old, he had had Epstein-Barr virus. At that time, he was tired, did not eat well and had stomach discomfort. He recovered after six to seven months. He also had a history of a milk allergy.

The pediatrician had tried a medication to reduce stomach acid which had helped a little.

Overall, Mark was fatigued enough that he was missing school and was not playing on the basketball team.

Mark started the 4 Stages diet and nystatin. Five weeks later, he reported that his energy had come back all the way. His sinus infection had resolved. He was staying in school. He was doing all his schoolwork. He reported that his general clarity of mind and his interest in activities had improved quickly and greatly.

Mark had no more abdominal pain at night. His sleep was fine. His appetite was much better. He had gained back some of the weight he had lost. He was now playing basketball again.

Mark still had some intermittent stomach discomfort and an occasional runny nose. He thought maybe some of the stomach discomfort occurred after drinking milk. He had been allergic to milk as a baby.

I saw him a year later in a store. He looked great and told me he felt fine.

What to do if your child seems to have an ear infection?

I have found that once people are treated for Candida, any ear infections which occur are less severe and usually do not require antibiotics. Most ear infections are viral and do not require antibiotics. Still there may be times that an ear may look a little infected to a doctor. If antibiotics are recommended, please give nystatin at the same time. The nystatin will decrease the yeast buildup caused by antibiotics. If the child is already taking nystatin, you may wish to increase the nystatin dose while the antibiotics is given.

One other option to consider is using homeopathy. Homeopathy is an alternative medical system which has no toxic side effects and can be very effective for numerous problems, including ear infections. I will explain a little bit about homeopathy here. However, a full discussion is a subject for a different book.

Homeopathy comes from the word "homeo" which means same or alike, and "pathos," which means suffering. Homeopathy is a form of medicine which was discovered and developed by a doctor in Europe named Samuel Hahnemann, starting in about the year 1800. This doctor was experimenting with quinine, a treatment for malaria. He took quinine and found that in him, a healthy person, quinine caused fevers. He thought this result to be very interesting because malaria is a disease of fevers. He restated a law that other medical philosophers had hinted at before, "like cures like." That which causes a symptom in a healthy person, can knock that same symptom out of someone sick with that symptom. This concept is totally opposite to the concepts of western medicine. Western medicine is also called allopathy, from "allo" meaning other or different, and "pathos," meaning suffering.

Dr. Hahnemann took his research further. He diluted, shook, rediluted and re-shook the quinine. He was hoping to get rid of the bad parts of the quinine and preserve the good parts. He found that not only did he get rid of the bad parts, but the resulting dilutions were actually more powerful than the original quinine. In fact, he diluted and shook the quinine so much that it seemed nothing was left of the original material. Yet, the mixture was still more powerful.

Dr. Hahnemann and others after him investigated other compounds found in nature. They found what symptoms these other compounds and natural materials might produce in healthy people. These investigations were called provings. Then the doctor would know which natural substances to use when treating a sick person. The doctor would match the symptoms of the sick person to the symptoms he or she knew the natural substances could produce in healthy people. Homeopathic doctors treated many disorders this way.

Homeopathy was very strong at one time in the United States. Homeopathic physicians were drafted on both sides in the Civil War. Such major medical schools as the University of Michigan were homeopathic medical schools. However, when antibiotics came out, homeopathic medical schools fell into disfavor. In the United States, all such schools closed or became allopathic medical schools. Very little homeopathy was taught.

Homeopathy has enjoyed a resurgence in recent years. Homeopathy is inexpensive and has no side effects. There are many excellent consumer-oriented books on homeopathy and how to use it in your home.

Homeopathy is challenging to learn because there is not just one remedy for a condition. For ear infections, there are at least seven main remedies. The right remedy depends on the exact symptoms. For example, in homeopathy, it matters whether the sore throat is worse on the right or left, is worse in the morning or night, and so on. In allopathic, or modern, medicine, these symptoms are not relevant. The challenge for the homeopathic physician is to match an individual's unique symptoms, including how they felt before the illness came on and how they feel during the illness, with the correct remedy.

For ear infections, homeopathy is most helpful in treating a cold so that the ear infection never starts. Once the ear begins to hurt, homeopathy offers many remedies, including Belladonna, Aconite, Pulsatilla, Chamomilla, Hepar Sulph, Mercurius and Silica. Different remedies are appropriate at different times. Many books describe these remedies and guide people in using them. I also caution people that they should be sure to consult with a homeopathic physician when using homeopathy, just as they should consult with an allopathic physician when using standard medicine. If they are lucky, they will find someone well versed in both systems who will know when to use which alternative.

This is, as I stated, a brief introduction to homeopathy. To go into any further discussion of homeopathic prescribing is beyond the scope of this book. I emphasize though that anti-yeast treatment complements both homeopathy and allopathic medicine. This is not an "either/or" choice. Medical consumers--you--should have the broadest range of effective treatments.

Notes

[1]When Candida albicans and Staphylococcus (a disease-causing bacteria) are grown together, C. albicans enhances growth of Staphylococcus (Virtanen, 1951). Candida albicans helps Lactobacilli grow (found in acidophilus) (Young et. al., 1951). Candida albicans may prolong the survival of C. diphtheria and Beta hemolytic streptococcus (the cause of "strep" infections) on dried swabs at room temperature (reported in Skinner and Fletcher, 1960). One investigator, Di Bella showed that the presence of Candida albicans reduced sensitivity of various bacteria to erythromycin and other antibiotics. C. albicans also contains a substance which enhances the virulence (disease-causing capability) of Pseudomonas aeruginosa and Proteus vulgaris (both disease-causing bacteria) (Yamabayashi, 1958). (Pseudomonas frequently affects cancer and other immunocompromised patients and Proteus can cause urinary tract infections, for example in men with prostate problems).

The relationship of Candida albicans and M. tuberculosis (the bug of tuberculosis) has also been investigated. Mankiewicz and her colleagues in a number of studies showed that portions of the yeast capsule are growth factors for M. tuberculosis (Mankiewicz, 1954, 1956, Mankiewicz et. al., 1959; Mankiewicz and Liivak, 1960). In living animals, both an extract of the C. albicans capsule and live C. albicans organisms aided the evolution of experimental tuberculosis in mice (Mankiewicz and Liivak, 1960). In other words, "do the yeast Candida albicans and Tuberculosis like each other and grow better together?" The answer is yes (Mankiewicz and Liivak, 1960). In summary, the yeast Candida albicans has been shown in a number of studies to help other disease causing organisms grow better.

Virtanen, I. Observations on the symbiosis of some fungi and bacteria. *Ann Med Exp. Fenn.*, 29:352-58, 1951; Young, G., Resca, H.G. and M.T. Sullivan. The yeast of the normal mouth and their relation to salivary activity. *J. Dent. Res.* 30:426-30, 1951; Skinner, C E., and D. W. Fletcher. A review of the genus Candida. *Bact. Rev.* 24:397-416, 1960; Di Bella, A. Costanza del fenomeno di interferenza sull' azione antibatterica dei nuovi antiotici in presenza di miceti. *Nuovi Ann. Ig.* 7:65-70, 1956; Yamabayashi, H. A zymosanlike substance extracted from Candida albicans. *Med. J. Osaka Univ.* 9:11-21, 1958; Mankiewicz, E., Mycobacterium tuberculosis and *Candida albicans* : a study of growth promoting factors. *Canad. J. Microbiology.* 1:85-89, 1954; Mankiewicz, E. Candida albicans. A means of detecting M. tuberculosis on culture media. Am. Rev. Tuberc. 75:836-40, 1957; Mankiewicz, E. and M. Liivak. The effect of *Candida albicans* on the evolution of experimental tuberculosis. *Nature* (London). 187:250-51, 1960; Mankiewicz, E., Stackiewicz, E. and M. Liivak. A polysaccharide isolated from *Candida albicans* as a growth promoting factor for *Mycobacterium tuberculosis. Canad. J. Microbiology.* 5:261-67, 1959; Mankiewicz, E. and M. Liivak. The effect of *Candida albicans* on the evolution of experimental tuberculosis. *Nature* (London). 187:250-51, 1960.

These references are found in *Candida albicans* by H. I. Winner and Rosalinde Hurley, J and A Churchill, London, 1964.

2 Ibid.

3Young, G., Resca, H.G. and M.T. Sullivan. The yeast of the normal mouth and their relation to salivary activity. *J. dent. Res.* 30:426-30, 1951.

Isenberg, H. D., Pisano, M. A., Carito, S. L., and J. I. Berkman. Factors Leading to Overt Monilial Disease. I. Preliminary Studies of the Ecological Relationship Between *Candida albicans* and Intestinal Bacteria. *Antibiotics and Chemotherapy.* 10(6): 353-363, 1960.

Chapter 14

Recurrent Vaginal Yeast Infections

Vaginal yeast infections are caused by the yeast Candida albicans. This yeast is found in the intestinal tract, in the mouth and in the vagina. Small amounts of this yeast in the vagina do not cause big problems, but when the amount of Candida increases, pain and inflammation result.

Recurrent vaginal yeast infections are a curse for many women, especially after they take antibiotics. Applying an antifungal cream may help temporarily, but the condition comes back. Why do such infections recur?

We must first know that the total amount of yeast in the gut affects the vaginal yeast. The amount of yeast in the vagina is only a part of the yeast in the body. If there is vaginal yeast, there is much more yeast in the gut. As we will see, the yeast in the gut can aid the vaginal yeast.

Antibiotics increase the amount of yeast in the gut as well as in the vagina

The amount of Candida in a woman increases significantly in the gut, or intestine, and in the vagina after she uses antibiotics. Many women have yeast infections only after antibiotic use, and view them as a "normal" (i.e., tolerable) side effect of antibiotics. In some women, the infections can come back even without antibiotics. These infections are itchy and painful.

What can be done about this yeast causing recurrent vaginal yeast infections? To answer this question, we must first look at the relationship of yeast found in various sites of the body.

Most of the yeast in the body is found in the intestinal tract

The yeast Candida albicans is found in the mouth and in the intestinal tract as well as in the female genital tract. The largest amount of yeast is found in the intestinal tract, or gut. Candida can be found anywhere in the gut. The gut is very long with a large amount of surface area. By comparison, the amount of surface area available to yeast in the mouth and female genital tract is smaller. The blood supply in the mouth and in the female genital tract is high. Usually that means that infections are less likely. More blood means more immune cells can get to the area, making infections less likely. So why do we find yeast at these sites?

Intestinal yeast helps vaginal yeast grow

Yeast in the intestinal tract is quite active. Recall what I presented in Chapter 8. Candida evades the body's immune system. Candida has a capsule which contains several factors to help Candida resist the body's immune system. The evidence suggests that Candida releases tiny portions of its capsule into the circulation. These pieces of capsule interact with immune cells in such a way as to suppress these cells and prevent them from reacting vigorously to Candida albicans. Releasing tiny portions of yeast capsule prevents the body's immune system from clearing Candida from the entire body, including the intestinal tract and female genital tract. In other words, immune cells which might attack the Candida in the female genital tract, instead are suppressed by the release of pieces of Candida capsule from intestinal yeast.

So even after an antifungal cream is applied to a vaginal yeast infection, the yeast which remains can grow back because Candida in the intestinal tract suppresses the body's immune response to such yeast. As long as Candida is present in the intestine and is making and releasing immunosuppressive factors, the body's immune system will have a difficult time clearing yeast from the vagina.

To clear vaginal yeast, you must first clear intestinal yeast

To clear yeast in the vagina, you must first clear the yeast in the intestinal tract.

In summary, the yeast which is in the intestinal tract makes growing yeast in other parts of the body easier. These other sites in the body include the female genital tract. So eliminating yeast from the intestinal tract is necessary to reduce the yeast in the vagina.

The way to stop this process is to clear the Candida out of the intestinal tract, using the treatment plan we give you in Chapters 16 and 17.

Cases of chronic vaginal yeast infections

Marilyn

Marilyn, 43, had had chronic vaginal yeast infections for 14 years. She had tried Gentian violet, boric acid suppositories (older remedies for vaginal yeast infections), and Nizoral pills (a systemic anti-yeast medicine). She had been tested for diabetes. Now she was using Monistat (an anti-yeast cream). Her worst yeast infections were prior to her periods. She had other problems. She was getting bronchitis about three to four times per year for which she was taking antibiotics. She had decreased energy. She was taking an antidepressant. She had stress headaches when working. She had problems with dairy products. She had a history of three years use of oral contraceptive pills.

She started the 4 Stages diet and nystatin. Nystatin powder was given both orally and in vaginal capsules. She came back four weeks later. She told me she felt much better. She had more energy and was sleeping better. The symptoms of vaginal yeast infection were gone. She had had some mild vaginal symptoms after stopping intravaginal nystatin capsules. So she had restarted them. I suggested she continue the anti-yeast diet and nystatin.

Thelma

Thelma, 35, told me she had chronic fatigue with poor memory and chronic feelings of being tired. She stated that if she tried to exercise her face would become red. Then she would feel hot and have to stop. She complained of chronic vaginal itching. She also had severe abdominal bloating about 1 to 2 times per week without pain. She would get migraines about once a week. She was fatigued since her first child was born eight years previously. As a teen she had slept a lot. She had had vaginal problems for some length of time. She had had memory problems for six years. Migraines had been worse in the last year. She had had migraines since she was a teenager. She had been diagnosed with asthma in the last year for which she received prednisone and antibiotics. She had a history of heavy long periods. She had taken oral contraceptive pills for two years but did not like the way she felt on them. She had been pregnant twice and had two children. She had had Hashimoto's thyroid disease five years before and was taking Synthroid (replacement thyroid hormone). She had been taking Prozac, an antidepressant. She was taking acidophilus and caprylic acid when she was first seen. Acidophilus and caprylic acid are both over the counter remedies for yeast.

Thelma started the 4 Stages diet and nystatin. She returned eight weeks later. She told me that she was better. Her chronic vaginal itching was gone. She was no longer taking naps. Her periods were now 28 days apart but still eight days long. Her memory was better. She had had two skin lumps since being a teenager and both were resolving. Her skin was not as dry. She stated that she had more energy. Her abdominal bloating had diminished. She had had only one migraine in two months.

Joy

Joy, 20, told me she had had two years of vaginal yeast infection. She had tried all the cremes and even Gentian violet. Two of the cremes had caused allergic reactions with extreme inflammation and bleeding. She also had sinus congestion and decreased energy. She had been treated for acne with the antibiotic tetracycline for years. She had received ampicillin for a cold prior to her first yeast infection. She had been on oral contraceptive pills for five years.

She started the 4 Stages diet and nystatin powder both orally and in intravaginal capsules.

She came back six weeks later and said that she was "OK." The vaginal nystatin capsules were working well. She had occasional discharge, but without external symptoms. She was using the capsules twice a day. When she dropped the capsule dose to once per day, her symptoms returned. Her sinus congestion was better and she was getting fewer headaches. Her energy was good. She was on the diet and was taking three eighths of a teaspoon of nystatin four times a day.

Ellen

Ellen, 44, told me she had had continuous vaginal yeast infections since the age of 30. At the time, she had a vaginal discharge. She was on oral contraceptive pills from the age of 24 to 34. She had had antibiotics for respiratory problems for 14 years. She had a nasal cyst removed about eight months previously. Since then Ellen had noticed increased drainage and felt as though she was choking. Homeopathic remedies had helped. She was off coffee, Pepsi, dairy products and was trying to get off sugar. She said that she was a "sugarholic." She had swollen glands as an early teenager and had facial acne and had received tetracycline. She had taken antibiotics all through life. In addition, Ellen had abdominal distention. She had a desire to urinate constantly. She had been treated for depression in the past. At the time of the appointment, her mental out look was all right. She also had hiatal hernia and throat spasms.

Ellen started the 4 Stages diet and nystatin. She came back about four weeks later. She had had a pinkish non-painful vaginal discharge for about ten days. She still had some residual vaginal discharge, but this was a major improvement. She was coping despite immense stress. She stated that others had noticed that she looked healthy. She felt rested when she awoke in the morning. Her nails were stronger. Her urinary urgency was gone. She complained of some gas after eating. She had abdominal pain after eating vinegar. The mucus was easier to clear from her throat although she said her throat was not better. Her sinus pressure was decreased. She was taking one quarter teaspoon of nystatin four times a day.

Conclusion

Treating intestinal yeast systemically resolves chronic vaginal yeast infections. There is no need to continue to suffer the itching and irritation these infections cause.

Chapter 15

Hormonal Problems Related to Candida

- *Premenstrual syndrome*
- *Endometriosis*
- *Decreased sex drive*

One of the health concerns my adult patients repeatedly raise is that they feel so lousy they do not have as much desire for sex as they used to. Or, without telling me this when they first come in, my patients will report back that their sex drive has improved after anti-yeast treatment.

Some of the health problems most responsive to anti-yeast treatment are decreased sex drive and other hormonal problems such as chronic vaginal yeast infections, endometriosis, premenstrual syndrome, and recurrent miscarriage. I discussed chronic vaginal yeast infections separately, in Chapter 14.

In general, I have found in my clinical practice that women display symptoms of Candida more than men do. One reason for this is that progesterone, the hormone of pregnancy and one of the main hormones in the birth control pill, somehow favors yeast growth. One study found that one of the main immune cells has more difficulty killing Candida when more progesterone is present.[1]

Progesterone increases in the second half of a woman's menstrual cycle, that is in the 10 to 14 days prior to menstruation, commonly referred to as the premenstrual period. Woman during this time can experience a number of symptoms including depression and headaches. One reason for the symptoms of premenstrual syndrome is the amount of yeast may rise with the increasing progesterone level. When we can decrease the amount of yeast in the body, premenstrual symptoms improve.

At least two other problems can occur due to yeast-induced sex hormone problems. These are endometriosis and recurrent miscarriage. Endometriosis is a painful condition in which the uterine lining grows in the wrong places. These conditions improve with the treatment of yeast.

The question is why we see these results from treating yeast. I suggest from what we know about yeast that two things are happening. First, let's go back to our discussion about immunology from Chapter 8. When I discussed how yeast defeats your immune system, I talked about how yeast puts out human hormone receptors on its surface. The receptors which have been studied are mainly those having to do with the immune system itself. However, other human hormone receptors may be present on the yeast capsule. If the human hormones are released, then circulate but bind to the Candida, as does complement (an immune molecule), then these hormones would do absolutely nothing. Yet hormone test levels would appear the same, because the hormones would still circulate in the blood.

Remember that the yeast receptors mimicking human immune cell receptors bind immune molecules such as complement, locking them up and taking them out of circulation. The same could occur with human sex hormones.

In other words, the patient would feel like the hormones were not present, and not doing their job, if the hormones were bound to the yeast. Killing and removing the yeast enables the hormones to bind to the body sites for which they are intended.

Another reason for improvement in functions related to hormones goes back to our discussion of toxic yeast chemicals in Chapter 2. The same yeast chemicals that sedate the brain may interfere with hormonal function, and/or the brain centers that control sexual activity. Removing the yeast removes these chemicals, and the patient functions more normally.

Whatever the mechanism may be for yeast to cause reproductive hormonal problems, you can see in the following cases how treating for yeast improves these problems.

Cases

Lashana

Lashana was 39 years old when she called me to tell me that she was trying to have a second baby and she had had two late miscarriages. She did not know what to do. Lashana did not have any medical problems and there was no reason for her to have late miscarriages. I prescribed nystatin for her and told her to follow the anti-yeast diet. She became pregnant within a few months and carried the baby to term.

The following case is typical. Cheryl came in with a multitude of problems related to yeast. She, like many patients, did not mention reduced sex drive as a problem. However, it was a problem. While treating Cheryl's other problems, her sex drive improved.

Cheryl

Cheryl, 35, complained of allergies and chronic sinusitis. She had sinus congestion, itchy eyes and throat. Sinus congestion caused headaches. She also had a four year history of knee pain and sometimes hand pain. She had had fatigue for five years. Allergy shots had helped until one year previously when she had moved. She had headaches. An antibiotic would cause a vaginal yeast infection. She had taken antibiotics for bronchitis for 15 years. Her menstrual cycles were normal. She stated that she had inhalant problems year round. She had been pregnant two times.

Cheryl started the 4 Stages diet and nystatin. She came back five weeks later. She stated that she was now having almost no headaches. She had more energy and more sex drive (not mentioned as a problem on the first visit). She still had intermittent sinus congestion but overall sinuses were better. Her knee pain was gone. She still had some postnasal drip and her itchy eyes were a little better and her throat still itched some. Her allergy shots seemed to work better now.

Ed

Ed, 58, came in because he had sinus problems, achy joints and arms, and colon spasms. Colonoscopy, a test to determine if something was wrong with his colon, showed nothing wrong. He was taking medication for the colon spasms. He was having pain in the lower part of his abdomen on both sides. His sex drive was weak. He had taken testosterone shots for years and then became discouraged and he was now taking testosterone pills. His sex drive was better on the shots than on the pills. Eight years previously he had found he had food allergies, so Ed was taking allergy shots for those allergies as well as for inhalants, pollens and molds. Ed got severe headaches if he stopped the allergy shots. He had skin problems on his hands. He had had intestinal mucus noted eight years previously. He had taken antibiotics intermittently. He was taking a number of medications including testosterone.

I started Ed on the 4 Stages diet and nystatin. He came back seven weeks later and reported that his sex drive had improved. His allergy shots were more effective and needed less often. He was tolerating more foods. His joints were no longer hurting. He had only occasional colon problems. He had not been having headaches.   He was taking one half teaspoon of nystatin four times per day.

Five months later, Ed reported that his sex drive had continued to improve. He was tolerating foods without getting pains and cramping. He was better able to tolerate milk and wheat. His sinuses had been clear until a week previously when they flared up. He had not been having headaches and he seldom had joint pains.

You might remember the following two cases from Chapter 3 on headaches. Both Norma and Carol came to me with many symptoms, including headaches and premenstrual syndrome. Carol also suffered from food cravings and addictions, so you met her again in Chapter 7.

Norma

Norma, 37, told me she felt emotional and cried easily, worse before her menstrual period. These symptoms began prior to her marriage 16 years earlier when she started the oral contraceptive pill, which she had taken for nine months. She then had two children. She had headaches every day, but a tooth retainer had reduced the headaches to 1 to 2 per week. She had regular menstrual cycles with heavy bleeding. She had received numerous antibiotics in the past, more frequently since being on the pill.

After seeing me, she started the 4 Stages diet and began taking a half teaspoon of nystatin four times per day.

She came back three weeks later and said that her headaches were gone. Her last period was better. She could deal with work better.

She came back after seven weeks of treatment. Norma reported that everything was fine, but she said that if she missed a few nystatin doses she started to feel bad.

Carol

Carol, 43, came to see me because she had suffered from premenstrual syndrome, extreme headaches, a plugged up nose and breathing troubles. This had gone on for the past 6 years. Allergy testing showed a few fall allergies. The sinus problems were so bad she had had sinus surgery the year before. When Carol came to me she was getting nosebleeds for a week before her period. Her nose was swelling up, especially at the incision sites. She still had headaches. She had a hiatal hernia, so she had been told to avoid coffee and chocolate before her period. Carol had intense chocolate craving during this time. She had a stressful job and could not concentrate. Concentration was worse before her period. She also had bloating and gas premenstrually. She had patches of dry skin on her scalp and face. Her skin itched before her period. She had no nasal congestion or headaches at other times of the month. Her periods were only 21 to 28 days apart, so she experienced these symptoms every few weeks. Carol had taken many antibiotics the year before the problem started. She had three children.

After discussing her problems and symptoms, I prescribed the 4 Stages diet for Carol. Two and a half months after being on the diet and taking nystatin, Carol came back feeling "pretty good.". Her headaches had decreased from being debilitating for days at a time, to only the day before her period. They were much less intense. She had more energy. Her nose was still plugging up and bled some, but less. The itching on her head was gone. The dry patches of skin on her ears and eyebrows were gone but the dry patch on her scalp was still present. Her bloating was gone. Besides all these physical symptoms improving, she craved chocolate much less.

Notes

[1]Nohmi and colleagues (1995) looked at whether neutrophils from estradiol treated mice were as good at killing Candida in the presence of progesterone, the main hormone of pregnancy. They found that the neutrophils did not kill Candida as well and they suggest that these hormonal effects are one reason why women are predisposed to vaginal candidiasis during pregnancy. Danazol also inhibited the killing of Candida.

Nohmi, T., Abe, S., Dobashi, K., Tansho, S., and H. Yamaguchi. Suppression of Anti-*Candida* activity of Murine neutrophils by Progesterone *in Vitro*: A possible Mechanism in Pregnant Women's Vulnerability to Vaginal Candidiasis. *Microbiol. Immunol.* 39(6):405-409, 1995.

PART IV
Treatment

Chapter 16

The 4 Stages Diet: How to Eat Better to Change Your Life

Healing yourself requires treating the yeast inside and preventing more yeast from entering your body

To this point, you have been reading about all of the problems yeast can cause, how yeast can hurt you and make you sick. You have learned that yeast make chemicals that can harm your health. You have learned that yeast can interfere with your immune system. You also have learned that certain foods can harm your health because they either contain chemicals similar to yeast chemicals, or they contain yeast growth factors. Those foods, then, are foods that can harm you health.

All of this information would be discouraging if you did not know you could get rid of the yeast. But you can! This chapter and Chapter 17 tell you exactly how to get rid of the yeast and heal yourself. This chapter is about what you eat--

how to keep more yeast from entering your body. Chapter 17 tells you how to destroy the yeast in your body.

The most important principle to understand is that healing comes from inside your body, not from the outside. What do I mean by this? Most people think of healing as defeating health problems by killing them or cutting them out. Most people would like to take a pill or have an operation to get rid of the yeast.

Solving the yeast problem requires a little of that, taking nystatin to kill the yeast inside your body, but *only* taking nystatin is ineffective unless you also prevent the yeast from returning. The only way to keep the yeast from coming back is to change what you eat. Yeast enters your body in food, and in foods that make yeast easier to grow. If you do not change what you eat, your problems will return.

The first, and most important step you can take right now is changing your diet. Nystatin does not work well without changing your diet. I have had patients who have been able to change their diet and get good results without taking nystatin. I also have had patients who have taken nystatin without changing diet, and have not gotten results.

Now let's pause for just a moment to answer the question you are dreading to ask: "Do I really need to do this for the rest of my life?"

For many people, the answer is **NO!** *About 80% of people only need Stages I and II to feel great* .

That's right. Most the people who use the 4 Stages diet report that they only need Stages I and II. In fact, many people benefit so much from eliminating just malt and vinegar on Stage I-A that they never go further. And that's OK.

For most people, going on Stage I of the 4 Stages diet, sometimes only for a few weeks or months, is enough. Try this. If you notice the symptoms coming back, you start the diet again. For other people, any deviation from the 4 Stages diet brings drastic consequences. So they need to adhere to the diet more strictly. Only you will be able to tell whether you are healthy on Stage I or whether you need Stage IV, or somewhere in between. That is the beauty of the treatment. You, the medical consumer, can evaluate how effective the treatment is and can monitor your progress.

So to recap, healing really comes in two parts: first, you keep the yeast out of your body by changing the foods you eat. Second, you destroying the yeast inside your body by taking the nontoxic medication nystatin.

The first step: changing what you eat

My patients who change their diets experience lasting health benefits. You can too. The only side effect of changing your diet is inconvenience. In fact, the hardest part of the food transition will be realizing--painfully--how many of the foods you eat are hurting you. We are the first to recognize, though, that changing diet presents social and emotional difficulties for everyone, including family.

Food has social and emotional implications as well as nutritional value. To change diet, you need to have a good enough reason to override these social and emotional difficulties. What better reason than giving yourself or someone you love the opportunity to live a better life?

We at Wisconsin Institute of Nutrition devised a diet, the 4 Stages diet, as an easy way not only to diagnose how severe your yeast problem is, but to make the transition to a yeast free diet relatively painless.

You have read this book up until now. You are motivated because you see in yourself, or a loved one, a medical problem that has not responded to conventional treatment, or you do not like the conventional treatments offered. Whether that problem be psoriasis, autism, migraine, multiple sclerosis, fibromyalgia, or any of the other problems discussed in this book, all diminish the quality of life of the person suffering from the condition. And all can be helped.

Untreated or ineffectively treated illness is at least equally as inconvenient as cooking a few special meals. You already are convinced that you, or someone you love, is suffering. You do not want this suffering to continue. So you are motivated!

But how do you do this?

We first published the 4 Stages diet and described it thoroughly in *Feast Without Yeast: 4 Stages to Better Health* (1999: Wisconsin Institute of Nutrition). We have made a few minor changes in this book, but you still can rely on *Feast Without Yeast*. *Feast Without Yeast* explains in great detail exactly which foods to eliminate and when. It also is a cookbook to help you make that transition easier. This diet is easy to follow, and the recipes are easy to use. This we know from the comments from our readers. So we won't repeat all of the information here. We will summarize the information here, and tell you here what you *can't* eat and why. I also will give you the lists of what you *can* eat.

In *Feast Without Yeast*, we also give you more than 225 original recipes to implement the diet. Our new companion cookbook, *Extraordinary Foods for the Everyday Kitchen*, gives you even more completely yeast free, gluten free and casein free recipes, as well as menus. If you are serious about changing the foods you eat, and want even more detailed help and recipes, *Feast Without Yeast* and *Extraordinary Foods for the Everyday Kitchen* (2003: Wisconsin Institute of Nutrition). Both *Feast Without Yeast* and *Extraordinary Foods* are available from Wisconsin Institute of Nutrition as well as Amazon.com and local bookstores.

Don't dive right in! Change your diet gradually and slowly

Most people like to start a new experience by diving right in. You are excited; you are motivated; you are ready!

So what do we tell you? Put on the brakes! Slow down!

Why? ***Because we want you to stick with the program.*** Your health is much more important than a fad. This is not a fad diet. This is the first step to a healthy life. We want you to take that step, then the next, then the next. But we don't want you to fall down the stairs.

The most effective way to change diet, a major life change, is to implement the change slowly and gradually. You don't learn to swim by jumping off the diving board! You need to learn new habits, find new foods to eat, and different ways to cook, which takes time. The 4 Stages are based on this concept. Instead of plunging in head first, we want you to eliminate the worst foods for your health first. This is why we divide the foods into four Stages. The Stages are prioritized: the most important foods to eliminate are in Stage I (A and B), the next most important foods in Stage II, the next in Stage III, and the final foods in Stage IV.

Not only do we divide foods into four different Stages, but we prioritize foods within each list of foods, or each Stage. This means that the most important foods to eliminate (the worst foods for you) are at the top of the lists. So, if you look at the list for Stage I-A, you would start with vinegar and malt, then work your way down the list. You would not start at the bottom of the list with cottonseed products.

All of those foods on Stage I-A are more harmful than the foods at the top of Stage I-B. So finish Stage I-A, assess whether you need to move on (which we discuss later), then go to the top of Stage I-B. This makes the diet very systematic and easier to follow.

Remember, also, that the lists are cumulative. Once you have completed Stage I-A, you keep those foods out of your diet. Every new food you eliminate is added to the other foods. Stage IV, then, really is Stages I-A, I-B, II, III and IV together.

The diets described in other anti-yeast books for the most part do not prioritize, so people trying to use those diets face the overwhelming task of trying to decide how to eliminate all of the foods at once. Those diets are much harder to follow.

A key point to remember is that **you may not need to eliminate all of the foods listed in the 4 Stages**. For many people, excluding the foods on Stage I is plenty. Many conditions respond quickly to minor changes. Depression, for example, usually responds very quickly to minimal dietary changes and taking nystatin. A woman in her late 30's who is fatigued after having several pregnancie, may only need to exclude the foods on Stage 1 and take nystatin for a few months. Children who have ADHD may need only to eliminate vinegar and malt. On the other hand, children who suffer from autism usually need to go at least through Stage III, and often through Stage IV, to gain maximum benefit from the anti-yeast treatment.

The picture I am painting for you should become clearer now. **You are the person who will determine which foods you need to avoid, based on how you feel.** Nobody knows you better than yourself! Because it's hard to remember exactly how you felt before you started the diet and at any given point, we suggest that you assess your health and make notes about what you are doing each week. I explain how to do that below, and give you a suggested format. You can photocopy the blank chart in Appendix B to use for this assessment, or you can devise your own strategy.

So if you start at the top of Stage I-A, with vinegar and malt (see below for why), and feel great after a week or two, stop there. Just continue eliminating vinegar and malt. If you do not feel as great as you want to feel, then move on to the rest of the foods in Stage I-A. Again, remember that the worst foods are at the top of the list. So if you want to get the most impact for your changes, eliminate food number 3 on the list. After some time, assess how you feel. If you don't feel as good as you think you should feel, go on to other foods. Then reassess. Eventually, you may proceed to Stage I-B, then Stage II, and so on. Stages I and II comprise the complete, basic yeast free diet.

Remember, about 80% of people only follow Stages I and II.

Stages III and IV are critical for some people

Stage III adds another dimension: eliminating gluten (a wheat protein) and casein (a milk protein). Many people who have problems with yeast, mold and fermented foods also have problems with gluten and casein. This is particularly true of people who have autism. I discuss this issue thoroughly in Chapter 4 on Autism. Again, you can try this stage to see how you feel. The fourth and final stage, Stage IV, is for those who really need it.

We urge you not to rush yourself to Stage IV too soon. Even if you love challenges and want to start with Stage IV, restrain yourself. Most people will not end up at Stage IV. I reported in *Feast Without Yeast* that my experience at the time was only a few people, maybe 5 out of every 100, need to go to Stage IV. This has remained true.

Stage IV is for particularly intractable problems, including Autism. Starting at Stage IV will deprive you of many foods that you may be able to eat. You don't want to lose that opportunity! In addition, going from a "normal" American diet to Stage IV is too drastic for most families. You will end up failing because you simply cannot enforce the diet. You will have much more success implementing the diet over the course of several months in a way that enables you to stick with it.

What to expect from dietary change alone

How much can you expect from dietary change alone if you cannot find a doctor who will prescribe nystatin? I usually tell people that changing your diet reduces yeast in the intestine and reduces the load of toxic chemicals, so much improvement results from dietary change alone. However, nystatin also adds a lot. I tell people about one third of the improvement is from diet and nystatin adds the other two thirds.

The basic principles of the 4 Stages diet

The 4 Stages diet is about better health. By taking out of your diet foods that are bad for your health, and introducing foods that are good for your health, you will experience positive changes. For example, adding safflower oil and sea salt is good for your health. The 4 Stages diet is low in animal protein, so some added fat is necessary to make you feel full and to flavor the food. Safflower oil is good for you in a positive way, containing all of the essential fatty acids. Sea salt is a good flavor enhancer and contains lots of minerals. Unless you know that salt causes you adverse health effects, use the salt in these recipes.

Children who have been following Stage IV, even with all of the fried foods, tend to be healthy and not have weight problems, because the overall diet is low in fat and healthy. You have a broad range of choice of recipes and foods to find combinations that satisfy your family.

I have one last word of caution before you start. Do not add back into your diet any food that you know already causes you problems! This seems pretty self evident, but I have had more than one patient who looked at the list for Stage I-A, concluded that these were the *only* foods bad for their health, and reintroduced aged cheese or some other food that they knew from previous experience caused major health problems. They then concluded that the 4 Stages didn't work! Don't fall into this trap!

Before you start, learn our rules for winning the anti-yeast game on page 281! Then read the overview for your plan for changing your diet. This starts on page 282. Having the overview in mind, you will be able to follow the specific plan much more easily. Then dive into changing how you eat, and how you feel.

WIN's Rules for Winning the Game!

1. Do only what you need to do. Or, don't be a martyr!

2. Think positively. You are not eliminating foods, you are eliminating a life of pain and difficulty.

3. Pay attention to your own observations. You are the best judge of how effective this treatment is, not your doctor, not your pediatrician, not your mother-in-law.

4. Accept your own mistakes. They will happen.

5. Accept your child's mistakes. Learn from them. Don't punish.

6. Expect to "plateau." You think you can make no more improvement doing what you're doing. Then move on to the next food or stage and see if that helps.

7. Keep trying to eliminate more foods until you are positive that you and your child can no longer benefit from any more dietary changes.

8. Beware the helpful saboteurs. These are the people who feel so sorry for your child, they give her chocolate candy or a brownie or cookie for a treat when you are trying to help her by eliminating those foods.

9. Expect gradual changes, not overnight miracles. Miracles happen very slowly.

10. Be organized and prepared for situations that involve food, like restaurants, travel and parties.

11. Garbage in, garbage out. Meaning: good food in, good feeling out.

12. Stick with what works, but be open to reevaluating.

13. Make dietary change a family project, not an individual chore.

Overview: transforming your life by changing your diet

You want to do it, you know you need to do it, but are overwhelmed and just need to get started! This is a basic step-by-step plan in how to change your diet, and your life in just 7 weeks. That's about the minimum time we suggest for making these changes. Of course, as we advise in the rest of this chapter, you do not need to go all the way through Stage IV if your health issues are resolved earlier! This overview is designed for the most sensitive people. But read through it so you have a game plan for how to proceed.

Week 1: Eliminate Malt and Vinegar Only

Day 1: Clean out your pantry, refrigerator and other food storage places

---Read all labels. Remove anything with "malt" in the label, including barley malt, malted barley flour, maltodextrin.

---Remove any foods containing vinegar, including salad dressings, condiments, breads, any other food

---If open, throw away (if you can't bring yourself to do this, then at least put them in a bag or box. You'll throw them away later.)

---If unopened, return to the store.

--Go shopping. Allow plenty of time. Use the shopping lists. Read the labels on each item, and buy only foods without malt and vinegar.

-- Start your chart of actions, behaviors and feelings (see example in a few pages, and Appendix B.) This chart will help you assess where you started and where you are going.

Days 2-7: Allow yourself to get used to a malt free, vinegar free diet. Try new recipes.

Week 2: Assess and Move Along

At the beginning of your second week, take a few moments to fill in your chart, assessing where you've been and where you are going. *If you are 100% happy with the changes in your life, and do not feel you need to do any more, great! Stop here.* If not, choose 1 or 2 more foods from the Stage I, such as chocolate, alcoholic beverages, peanut butter, nuts or aged cheese.

During this week, try eliminating one of two foods. Then assess how you are doing.

Weeks 3-6: Experiment, Evaluate, and Eliminate

You are the best judge of how things are going. Take the time at the beginning of each week to fill in the chart, and assess whether you need to continue. If you do, gradually eliminate the foods on the lists in Stages I and II. You may "plateau", but still may not be 100% satisfied with the results. This is a sign you need to eliminate other foods. Another reaction you may have is that as you eliminate some foods that bother your body, other foods that still bother your body may have an even more extreme reaction. Don't give up! Keep going.

At time point, you can start nystatin, a nontoxic anti-yeast medication. This will kill the yeast inside your body. Usually improvement is seen within a day or two of starting nystatin and being on a yeast free diet. See the information later in Chapter 17.

Week 7: Eliminate Gluten (Wheat) and Casein (Milk protein)

By now you have eliminated the foods on Stages I-II. At this point, you may want to go on to Stage III, eliminating casein and gluten. Many people may start out on a gluten/casein free diet and add the yeast free part. If you are not already on a gluten/casein free diet, and are not feeling 100% great, then try this. Some estimate that about 70% of people with autism benefit from a gluten/casein free diet. Other conditions also respond well.

Day 1 of week 7: Find and eliminate all foods containing gluten and casein:

> ---Gluten containing foods include wheat, barley, oats, rye.

> ---Some foods may contain "hidden" gluten because they are on mixed production lines. If you think you have eliminated gluten, but still are having problems, call the manufacturer to determine if the foods have gluten.

> ---All cow's milk containing foods have casein. Many people tolerate goat's milk who do not tolerate cow's milk casein. Check labels on soy cheese and other products for "caseinate," because this is the same milk protein that causes problems. Many people who generally are intolerant of casein can tolerate butter, because the amount of protein in butter is minimal. You will need to experiment by taking out a certain food for at least a week, then adding back to see if there are health changes.

---Check the labels on all foods labelled "gluten/casein free." Many are not yeast free and contain chocolate, vinegar, malt, peanuts, and other foods that will set you back.

After eliminating foods that have gluten and casein, go shopping. Buy Rice Pasta. We recommend Pastariso brand, because we like the taste and texture, but there are other brands you may enjoy. Make as many great tasting foods as you can so your family doesn't mind the change.

Weeks 8-12: Experiment and Evaluate

This again is a time for you to evaluate how things are going. If you would like to do so, try adding back a food that you question whether it is problematic. Then watch for the next few days to see if there are any changes. Changes may be behavioral. They may take the form of rashes, eczema or hives, stomach problems, or other types of changes. Watch and keep track. Beware that the effects of gluten may take months to leave the body's system, so experimenting with wheat right now may not give you a good idea about what's going on.

Week 13: Stage IV

By now, you've been on this plan for 3 months. You should see changes. If you believe you need to go further, now is the time to try Stage IV. If you still are suffering from skin problems (rashes, eczema or hives) you should try eliminating eggs first, then moving on to Stage IV.

The Details: week one--day one

You have read the overview and understand generally how the 4 Stages diet works. But now you want to know how to implement the plan day by day. Here is where it starts: week one, day one. Today, you will do five things:

1. **Start a chart or a record, so you can keep track of how you are feeling now and how you feel as you eliminate foods from your diet and add nystatin. Keep notes, because you will want to go back to them, especially if you run into people who do not believe you and convince you not to believe yourself.**

2. **Go through your cabinets, refrigerator and pantry looking for foods with vinegar and malt.**

3. **Go shopping to find foods without vinegar and malt.**

4. **Cook a great meal with at least one new food.**

5. **Enjoy eating that new food!**

Eliminating Vinegar and Malt

I described in Chapter 2 the worst food enemies, which are vinegar and malt. So it makes sense that you would eliminate these two foods first. You will see that these are the first two foods on Stage I-A. The first week, you *only* eliminate these two most important foods.

On the first day, go through your cabinets, refrigerator, and pantry to find which foods have vinegar and malt.

Why these two? Because, as I explained in Chapter 2, vinegar and malt both contain toxic yeast chemicals. There are 20 bad chemicals in malt. No wonder kids can't concentrate when they eat malt-filled breakfast cereals in the morning! No wonder you have a hard time staying awake during meetings

Sample Chart for Assessing your Health

Date:	Health/ Behavior	Actions taken:	Other changes:
1/1	tired, frustrated, low tolerance for noise, rash, itchy, poor sleep, aggressive, hyperactive	started Stage I; eliminated vinegar and malt	none, still on winter break
1/8	less tired, sleep better, not aggressive, said "hi" to dad	stayed with removing vinegar and malt. This is getting easier. I found some great pretzels that we love, also pizza sauce	back in school.
1/16	rash starting to improve, sleep better. Finally slept the whole night! School going ok. Teacher noticed some changes, said J. quieter in class.	Removed soy sauce.	Therapist quit; need to find a new one.
1/23	Incredible week! great behavior, slept well.	Started nystatin, eliminated chocolate	
1/30	bit a kid at school. Found out he had chocolate chip cookies and pickles for a treat at lunch! Talked to the teacher about the diet again.	eliminated chocolate for sure, thinking about cheese	New therapist working out well.

after eating a bagel! Also, malt is a growth factor for yeast, meaning that malt helps yeast grow. Vinegar is no better. Vinegar also contains numerous toxic chemicals. We use vinegar to clean our floors and bathrooms. Just imagine what this does for your insides!

Surprisingly, bread yeast is not on the "most wanted" list of health enemies. The yeast from bread has only a short time to make toxic chemicals and is killed during the baking process. In addition, this is a different kind of yeast than the yeast in your body. Even white table sugar (sucrose) is not eliminated until Stage II. So at this point, just concentrate on vinegar and malt because, in my clinical experience, these two foods have the most immediate impact on health, and you will see the most immediate benefit from eliminating them.

You will start just by looking in your pantry or food cupboard for anything that has the word "malt" or "vinegar" in it in any form, including "barley malt," "malted barley flour," "maltodextrin," and any other form. Both vinegar and malt are ubiquitous in our modern, processed food supply. You will find malt in many cereals, crackers, breads, white flour, bagels and in many health food snacks. Although barley malt is the worst offender, other types of malt, including maltodextrin (usually malt added to corn syrup), also cause harm. When shopping read labels carefully and avoid anything with the word "malt" in it.

You will be amazed at how many commercially prepared foods, and even white flour, contain "malt," "barley malt," "maltodextrin," or "malted barley flour." So even baking your cookies "from scratch" does not guarantee health. I have found only one brand of commercial white flour that does *not* contain malt. Even some whole wheat commercial flours contain malt.

Malt is in most brands of bagels, for example. So on Stage I-A, you must eliminate bagels, unless you are lucky enough to find malt-free bagels. If you buy your bagels at a bakery, ask the baker to read the label on the bag of flour. We have a local baker, for example, who uses malt free flour. You need to read the labels on all flour packages.

STAGE I-A
REMOVE THE FOLLOWING FOODS FROM YOUR
DIET in this order of importance:

Barley malt and all malt products, including maltodextrin

Substitute: similar foods that do not contain malt. For example, most General Mills cereals, do not contain barley malt.

Vinegar

Substitute: freshly squeezed lemon juice; tomato paste for ketchup; see the many recipes in this cookbook for sauces and salad dressings.

After eliminating malt and vinegar, eliminate these foods:

Chocolate

Pickles and pickled food, including: herring, pickled tomatoes, pickled peppers

Alcoholic beverages and nonalcoholic beer

Aged cheese

aged cheeses include: cheddar, Swiss, parmesan, romano, blue cheese, roquefort and similar cheeses. **Substitute**: mozzarella; brick cheese; jack cheese. For some people, goat's milk or sheep's milk Feta cheese is acceptable, as long as the cheese is fresh and packed in water

Soy sauce and tamari (substitute: sea salt)

Fermented soy products such as tempeh and miso

Other fermented foods, such as rice syrup and brown rice syrup

Worcestershire sauce

Cottonseed oil and other cottonseed products

For the foods you CAN eat, see the next page. . . ...

Allowable Foods on Stage I-A
You can eat everything you now eat, except the specific foods eliminated.

Fresh and freshly frozen Meat, Fish and Poultry: all kinds (preferably hormone and antibiotic free)
♦ Canned Tuna Fish (labelled "very low sodium"--only ingredients Tuna Fish and Water)
♦ Processed meats, including hot dogs, bacon, salami, luncheon meats and bologna

Fresh Produce: all kinds, including all fruits and vegetables, fresh and dried herbs

Dry Goods:
♦ Dried beans (all types)
♦ Coffee and tea
♦ Rice and all rice products
♦ Unprocessed clover honey
♦ Flour (malt free unbleached white flour, whole wheat, or gluten free flour made from other grains)
♦ Dried and packaged cereals without malt
♦ Chips that are not cooked in cottonseed oil or peanut oil
♦ Dried fruit and raisins
♦ Maple syrup
♦ Nuts and peanuts

♦ Oils-all except cottonseed and peanut
♦ Oatmeal and oats
♦ Pasta (whole wheat, semolina or rice)
♦ Sea salt
♦ Soda drinks
♦ Snack foods such as cookies, pretzels, crackers, etc., that have no malt, chocolate or vinegar, or rice syrup
♦ Spices (cinnamon, etc.)
♦ Sugar
♦ Whole wheat tortillas or chapatis, matzah
♦ Whole wheat bread (malt free)
♦ Whole wheat flour

Milk, Butter and Eggs:
♦ Eggs, preferably hormone and antibiotic free
♦ Milk, all kinds
♦ Butter
♦ Cottage cheese, ricotta cheese, mozzarella and other non-aged cheeses
♦ Yogurt
♦ Ice Cream without chocolate or vanilla flavoring (try the new packaged sorbets!)
♦ Ice Cream substitutes (tofu or rice-based)

You also need to be aware of the malt in processed foods. One time we happened to be taking a walk when we saw cartons of food being delivered to a major fast food chain restaurant. We took the opportunity to read the ingredients on the box. One of the main ingredients in the hamburger buns was malt! If in doubt, ask.

You will be shocked at what has malt. One of my favorite examples is a spiced cider mix. Whoever would expect malt there? We also have found maltodextrin in ice cream.

You will be able to find substitutes for most of these foods if you spend time looking for them. Just allow yourself extra time at the store for the first week or two; gradually, you will get to know what foods you can buy. For example, if your favorite brand of pretzels or cereal contains malt, look at all the other brands. There usually is at least one comparable brand without malt. Some helpful hints: at the time of writing this book (that is, winter 2003), generally we have found that most General Mills cereals, such as Cheerios and Kix, do not contain malt. Some Quaker cereals do not contain malt. Some breads and bagels do not contain malt, but most do. You need to read the labels yourself! If you have doubts, most manufacturers list a toll-free number to call with consumer questions. I have called these numbers many times to be sure no hidden malt is in their products. They usually are very helpful when they understand a potentially bad reaction to their food is possible.

One word of caution: keep checking the labels from time to time. Manufacturers change ingredients frequently. We have had more than one favorite brand go out the window because the manufacturer added malt. We check the labels on all packaged products each time we purchase them. We frequently verify that our local bakery still does not use malt.

Vinegar is high on the list of health enemies. I explained in Chapter 2 how vinegar contains toxic yeast chemicals. So, again, go through your pantry, refrigerator and cabinets to get rid of the vinegar. You will find vinegar in virtually all condiments, including ketchup, mustard, sauces and salad dressings. You can find commercial substitutes for most products. For example, some people use a pizza sauce as a substitute for ketchup. You also can use canned tomato paste. If you have suspicious children, empty a ketchup bottle, refill it with slightly watered-down tomato paste, and squeeze. Even within brands, some sauces contain vinegar and others don't. So you need to check the ingredients on each and every item.

In *Feast Without Yeast,* we have recipes for alternatives to these condiments, as well as lists of common substitutes. We also post some recipes on our website, http://www.nutritioninstitute.com.

So the first few days may be a little surprising for you. You will need to look through your pantry and refrigerator to learn what foods you currently are eating that are causing you to feel ill. Most people are shocked when they do this. If throwing out food is difficult for you, put the food in a paper bag or a box and hide it. You don't have to throw it away, but don't eat it. After a few months, you'll be surprised to learn that you actually have survived without these foods. If you have unopened food, you may be able to return it or donate it to a food pantry. We have mixed feelings about donating potentially toxic food to food pantries, but this is an alternative to throwing it out or eating it yourself.

Week 2: assess your health and determine your next step

After you have followed the diet for about a week, eliminating vinegar and malt, we suggest you use the chart you made earlier to:

1. **Assess your health.**

2. **Make notes about your health and any other changes you made in addition to eliminating vinegar and malt.**

3. **Decide what you need to do to move on.**

4. **Move on to the rest of Stage I-A. Select one or two foods to eliminate this week.**

5. **Add safflower oil and brown rice to your diet, if you don't already use them.**

What do we mean by assessing your health? This means just to think about how you feel, emotionally and physically. You are the person who knows your health best. If you are a parent helping a child adjust to the 4 Stage diet, assess your child's health and behavior. Often behavior is a good indicator of health issues. Do you (or does your child) feel better? Do you have more energy? If you are working with a child, has their behavior improved? Has their eating improved? Do you see no changes at all? Record your thoughts on your progress chart.

If you are100% happy with where you are now, stop here. Congratulations! You have managed to solve a lifelong problem in a week. We are not being facetious. Many people have improved so much just by eliminating vinegar and malt that they need go no further.

If you feel you can improve more, continue with the other foods on Stage I-A.

Chocolate, aged cheese, soy sauce and the rest of Stage I-A

Now look at the remainder of the list for Stage I-A. You will find chocolate, soy sauce, pickled foods, and other foods.

Choose just one or two foods to eliminate this week. These are challenging foods for many people!

In Chapter 2, we explained why each of these foods can cause problems for you. I will just review the reasons here.

For many people, chocolate is the food that seems most difficult. Although they may consider chocolate one of the four basic food groups, they may be surprised to know that they can live without it. Some people may even be able to eat a little chocolate every once in awhile without suffering greatly. If you are one of those people, you will need to monitor your reactions closely.

Chocolate has two problems. Chocolate is dried and fermented with a fungus. Chocolate also contains a chemical compound which is similar to one of the yeast chemicals. Unfortunately, there is no substitute for chocolate. But we have many excellent dessert recipes that do not use chocolate.

Pickles, pickled foods, soy sauce, worcestershire sauce and alcoholic beverages are fermented. Aged cheeses are highly mold contaminated. At this point, we do not remove milk from the diet. However, it is worth noting that removing milk (as in Stage III), may be helpful because milk is a growth media for Candida.[1] The longer dairy products sit, the more time microorganisms have to work on them.

Aged cheese definitely is a problem. You should eliminate aged cheese on Stage I-A. Cheese processing starts with bacteria, and the starter bacteria for cheese makes the nerve poison hydrogen sulfide.[2] This means that aged cheese is worse than mild cheese, which is worse than cottage cheese and yogurt. So on Stage I-A, eliminate aged cheese such as swiss, cheddar, or any other cheeses that say "aged." Stick to the milder cheeses, like mozarella, farmer's cheese, jack cheese, and cottage cheese.

Some people may want to start limiting dairy products now, anticipating Stage III. Try substituting non-instant, non-fat milk powder for milk. This product is evaporated quickly, so nothing has time to grow in it. It tastes sweeter than regular milk, and is harder to mix than regular instant milk.

Cottonseed oil and cottonseed meal are problematic because the cottonseed plant is often mold contaminated and the products of the mold end up in the cottonseed oil.

Fermented products, including the soy products of tempeh, miso, tamari and soy sauce, also should be eliminated. Fermented rice products, including rice syrup and brown rice syrup should be eliminated. Other fermented products used in Asian cuisine, such as fermented black beans, fermented szechwan vegetables, also fall into this category. All of these foods can wreak havoc with your body. The best substitute for soy sauce is sea salt. If you cook heavily with these fermented foods, gradually change your cooking style. Increase the amount of sea salt while decreasing the fermented sauces and products. You will find that as you feel better, you don't miss the old foods. The primary cook in one family we know simply eliminated soy sauce without anyone ever noticing. Her family found that the food was even more delicious because they tasted the natural flavor of the meats and vegetables in the foods.

Potato chips are OK at this point, but check the types of oil in which they are cooked and check for black or green moldy spots on the chips. We suggest not eating a lot of potato chips.

A word about alcohol

This book assumes that you are not drinking alcoholic beverages, or at least not a lot of alcoholic beverages. All alcoholic beverages contain chemicals which kill bacteria and enable intestinal yeast to grow. If you drink alcoholic beverages, you will have more yeast. However, you do not have to stop drinking alcohol to do this diet. You should still start with vinegar and malt, and eliminate other foods as you go. Alcohol is on Stage I-A. Following the rest of this diet will help you cut your urge to drink. We discuss concepts about food addiction in Chapter 7; the same principles apply to addiction to alcohol. Eating malt, for example, will trigger your desire to drink alcohol. Stopping malt will help you stop drinking alcohol.

I suggest that you stop all drinking for at least a few days, then add back a small amount of alcohol. See how you feel, and decide what to do from there.

Increase the good foods in your diet

While you are eliminating foods, increase the amount of good food in your diet. Brown or whole grain rice is an exceptionally nutritious food. This is rice in its natural state, with the brown outer husk left on. White rice is brown rice without the outer hull. That outer hull contains fiber and nutrients. There are many varieties of brown rice, from long grain to short grain, from sweet to tart. There are other types of whole grain rice, including red rice and wild rice. Experiment and enjoy. If you are not used to eating whole grain rice, start by mixing a little brown rice with white rice, then gradually change the proportions. We give specific cooking instructions for basic brown rice in *Feast Without Yeast,* and have many delicious recipes using brown and other whole grain rice in both *Feast Without Yeast* and *Extraordinary Foods for the Everyday Kitchen..*

Week 3: Taking Stock

You have been on the 4 Stages diet for 2 weeks. You have eliminated malt, vinegar and one or two other foods. You now are starting to clear out the toxic wastes polluting your body.

1. **Assess where you are and what changes you see. Write them down on the chart.**

2. **Decide whether to just continue what you are doing, or which foods to eliminate next.**

3. **Look back at where you were when you started. Do you see any difference in how you feel?**

Now you need to go back to assess how you feel. What changes have you noticed in yourself? Your child? Get out the chart, and record your thoughts.

At this point, you should decide whether you want to just stay where you are for a few more weeks, or whether you want to try eliminating more foods. Look at the rest of the Stage I-A list and week by week, gradually eliminate more. Many people like the changes they see so much that they have the courage to eliminate a few foods, not just one or two. Make the decision for yourself.

You also have been on the diet long enough that you probably would benefit from taking nystatin, discussed in Chapter 17. Make an appointment with your health care provider to discuss this with him or her.

At the end of each week, make time to assess how you feel, record any changes that you have made or seen in yourself or your family, and look back at the chart to see how far you have come.

When it is time to do so (and only you will know this), move on to Stage I-B.

Stage I-B

Stage I-B calls for removing more of the extremely fermented and moldy foods from your diet. As you did with Stage I-A, remove the foods at the top of the list first, and only remove one or two at a time.

1. **Look at the Stage I-B list. Starting at the top of the list, eliminate those foods until you feel that you have reached the healthiest point you can.**

2. **Add more good foods, including potatoes, vegetables, new fruits, and beans.**

STAGE I-B LIST

REMOVE THESE FOODS FROM YOUR DIET:

Nuts and peanuts

Apples and apple products

Grapes and grape products

Coffee

Hot dogs, salami, bacon, luncheon meats and other processed meats containing sodium nitrates and/or sodium nitrites.

"Natural" hot dogs can still be eaten at this point.

Start limiting your intake of bread made with yeast

The first food group to remove is nuts and peanuts, including all products made from nuts and peanuts (including peanut butter). Nuts and peanuts are inherently mold contaminated. The most moldy peanut butter is natural (unprocessed) peanut butter. Many people also have peanut allergies.

The next food group for children covers apples and grapes and all products made from apples and grapes. Some adults can still handle grapes, however. Like other foods, I suggest eliminating them for awhile then adding them back to see if you react. Apples contain a natural antibiotic, which can promote yeast growth. Apples and grapes also contain yeast by-products that Dr. William Shaw, of the Great Plains Laboratory, has isolated. In clinical experience, apples, apple juice, grapes and grape juice wreak havoc in children sensitive to yeast. Substitute other fruits for apples and grapes. Pears substitute well for apples; fresh berries substitute well for grapes. You can experiment with this. Some children have terrible reactions to apple juice, but can tolerate peeled fresh apples, or fresh apples cooked into apple sauce.

Coffee should be removed at this point. Coffee contains some of the same chemicals as malt. For some people, this may be the most difficult food to remove. Work at it slowly and gradually. First cut down on the amount of coffee you drink. Substitute herbal tea or hot lemonade and experiment with the wide variety of flavors you can find.

Remove processed meats containing sodium nitrate and sodium nitrite. Sodium nitrite stabilizes the red color in processed meats and adds flavor. Sodium nitrate helps cure the meat and slowly breaks down into sodium nitrite. These additives have been linked to cancer. I see in my clinical practice many children who have behavioral difficulties as a direct result of eating meats with these additives.

Finally, start limiting your intake of bread made with yeast. For many people, this will be a dramatic change. Stage II totally removes yeast bread. Although the yeast in bread is not the same as the yeast in your body, it still makes bad chemicals that can hurt your health.

As before, go through this list slowly. Take one food at a time, find substitutes, experiment with other flavors, and become comfortable. This is a gradual process that will take time.

On Stage I-B, you should also add more good foods to your diet. By now, brown rice is a regular part of your diet. Concentrate on increasing potatoes and vegetables, substituting other fruit for apples and grapes, such as plums, berries, and pears, and getting your children used to drinking water or fresh orange juice instead of apple juice. Fresh lemonade made with honey is a great substitute for juice. We have delicious recipes for hot and cold lemonade in *Feast Without Yeast*.

Begin to add beans to your diet. Beans come in many sizes, flavors and colors, from standard beans such as lentils, kidney and garbanzo beans to exotic beans, such as Anasazi and "cow" beans. As you begin to eat more beans, you will find that you will like some more than others. Again, in *Feast Without Yeast* and *Extraordinary Foods for the Everyday Kitchen*, we provide detailed cooking instructions for beans, as well as several delicious recipes. You can introduce beans slowly by adding them to soups. You will be happy to know

Allowable Foods on Stage I-B

Fresh and freshly frozen Meat, Fish and Poultry: all kinds (preferably hormone and antibiotic free)
♦ Canned Tuna Fish (labelled "very low sodium"--only ingredients Tuna Fish and Water)
♦ "Natural" hot dogs and luncheon meats only; nothing with sodium nitrates or sodium nitrites

Fresh Produce:
♦ All kinds except apples and apple products, grapes and grape products. These include all fruits and vegetables, fresh and dried herbs
♦ Fruit juices

Dry Goods:
♦ Dried beans (all types)
♦ Tea
♦ Rice and all rice products
♦ Unprocessed clover honey
♦ Flour (malt free unbleached white flour, whole wheat, or gluten free flour made from other grains)
♦ Dried and packaged cereals without malt
♦ Potato chips that are not cooked in cottonseed oil or peanut oil

♦ Dried fruit and raisins
♦ Maple syrup
♦ Oatmeal and oat products
♦ **Oils**-all except cottonseed, peanut, and other nut oils
♦ Pasta (whole wheat, semolina or rice)
♦ Sea salt
♦ Spices (cinnamon, etc.)
♦ Soda drinks
♦ Snack foods that are free of malt, chocolate and vinegar, including cookies, pretzels, and crackers
♦ Sugar
♦ Whole wheat tortillas or chapatis
♦ Whole wheat bread (malt free)
♦ Whole wheat flour

Milk, Butter and Eggs:
♦ Eggs, preferably hormone and antibiotic free
♦ Milk, all kinds
♦ Butter
♦ Cottage cheese, ricotta cheese, mozzarella and other non-aged cheeses
♦ Ice Cream without chocolate or vanilla flavoring (try the new packaged sorbets!)
♦ Ice Cream substitutes (tofu or rice-based)

that adding beans to your diet, as well as other high fiber foods, will not necessarily cause excess digestive gas. The gas comes from the fiber in beans. The more fiber you eat, the more your body gets used to digesting it. Different cooking techniques also decrease the amount of digestive gas from beans.

Beans also are an excellent source of calcium. Beans form part of a complete protein when combined with rice. So introducing beans to your diet also establishes a base from which you can eliminate dairy in the future if you need to do so. We give you excellent recipes for beans in *Feast Without Yeast* and in *Extraordinary Foods for the Everyday Kitchen*.

Moving on: Stage II

By now you are accustomed to Stage I, which eliminates the most highly problematic foods. Again, evaluate how you feel. Take time to look back at your journal, fill in the important information, and assess whether you think you can feel better than you do now. If so, it is time for you to move on to Stage II. Some health care practitioners prefer to skip Stage II at this point and move directly to Stage III, eliminating wheat and milk, then going back to Stage II. Skipping to Stage III particularly makes sense for health conditions such as autism.

By now, you also may have started taking nystatin to see whether it helps you.

For Stage II, make sure that you are eating brown rice instead of white rice, lots of good potatoes, beans and vegetables. You should follow Stage II for about four to six weeks, continuing to take nystatin. If you are not taking nystatin, consider at this point starting it.

The first foods to remove from your diet on Stage II are baked goods containing yeast, including bread. These foods actually contain yeast, and at this point, removing yeast will help you.

Next, stop eating and cooking with corn and rye, which both are highly mold contaminated. Fermented foods to

Allowable Foods on Stage II:

Fresh and freshly frozen Meat, Fish and Poultry: all kinds (preferably hormone and antibiotic free). Limit your intake of chicken and turkey

♦ Canned Tuna Fish (labelled "very low sodium"--only ingredients Tuna Fish and Water)

♦ "Natural" hot dogs and luncheon meats only; nothing with sodium nitrates or sodium nitrites

Fresh Produce:

♦ all kinds except apples and apple products, grapes and grape products, mushrooms and bananas. These include all fruits and vegetables, fresh and dried herbs

Dry Goods:

♦ Dried beans (all types)

♦ Tea

♦ Brown rice and all brown rice products

♦ Unprocessed clover honey

♦ Flour (malt free unbleached white flour, whole wheat flour and gluten free flour made from other grains, except rye and corn)

♦ Dried and packaged cereals without malt

♦ Potato chips that are not cooked in cottonseed, corn or peanut oil

♦ Oatmeal and oat products

♦ Oils: expeller pressed safflower oil, canola oil and olive oil

♦ Pasta (whole wheat, semolina or rice)

♦ Sea salt

♦ Snack foods and other processed foods that are **free** of: NutraSweet (aspartame); monosodium glutamate (MSG), malt, rice syrup, chocolate, yeast, vanilla extract, corn, rye and vinegar, including cookies, pretzels, matzah and crackers

♦ Whole wheat tortillas or chapatis

♦ Whole wheat flour (malt free)

♦ Yeast free bread from *Feast Without Yeast*

Milk, Butter and Eggs:

♦ Eggs, preferably hormone and antibiotic free

♦ Milk, all kinds except buttermilk

♦ Butter

♦ Yogurt

♦ Cottage cheese, ricotta cheese, mozzarella and other non-aged cheeses

♦ Ice Cream and Ice Cream substitutes without chocolate or vanilla flavoring

STAGE II LIST
REMOVE THE FOLLOWING FOODS FROM YOUR DIET
in this order of importance:

Baked goods containing yeast, including bread.

  Substitute: yeast free bread

Corn and rye

Vanilla extract

Dried fruits and raisins

Concentrated fruit juice, especially fruit juice

concentrate used as a sweetener

Monosodium glutamate (MSG)

Aspartame (NutraSweet)

Maple syrup

Bananas

Cut back on meat and fish, especially on chicken and turkey

Colored spices (green herbs are OK)

  (Cinnamon, cumin, chili powder are colored spices. Basil, thyme, oregano, etc., are green herbs.)

Mushrooms

Soda drinks and Buttermilk

Cooking oils except safflower oil, olive oil, and canola oil

Table sugar (sucrose), including both white and brown.

  Substitute: unprocessed honey

Margarine

  margarine has a host of problems. The human body does not metabolize it. Butter, a natural product, is much better for the body, even though it contains cholesterol. Substitute: butter remove are: dried fruits and raisins, vanilla extract, concentrated fruit juice, and buttermilk.

Maple syrup should be removed. It contains some of the same toxic chemicals as vinegar, including acetol. Monosodium glutamate (MSG) and aspartame (brand name of NutraSweet) both are food additives that I have found, in my clinical experience, may cause headaches and other problems.

Stage II also calls for removing table sugar (sucrose) and substituting honey. Sugar gives yeast a boost but itself is not toxic. Eliminating foods such as malt is far more important than eliminating sugar. I note however, that soda contains a tremendous amount of sugar (about 12 teaspoons in a can!) and should be avoided. Eating a small amount of sugar occasionally is not going to bring the yeast back; eating a small amount of malt occasionally will cause major problems.

Dried fruits and raisins may be treated with sulfites which kill bacteria and help yeast. The problem with concentrated fruit juice is that there may have been moldy fruit used in producing the concentrated fruit juice. Honey is a more complex sugar than table sugar, and it does not grow mold. Colored spices are eliminated on Stage II, such as cinnamon, allspice, dried mustard, paprika, and so on, because many are inherently mold contaminated. We find, however, that you can eat something with some cinnamon or other spice every once in awhile. You need to experiment for yourself. The dried green herbs are fine, however. These include delicious flavors like basil, oregano, marjoram, thyme, rosemary and other herbs.

Stage II also eliminates and edible fungus, including mushrooms, as well as carbonated sodas, most of which have corn based products in them.

I advise not using margarine at this point, too. Margarine has a host of problems. The human body does not metabolize it. Studies have shown that the fats in margarine (trans-fatty acids) are as bad for the body as highly saturated fats. Butter, a natural product, is much better for the body, even though it contains cholesterol. If you are trying to eliminate saturated fats or cholesterol, substitute oil for margarine. I recommend expeller pressed safflower oil. Canola and olive oil are acceptable as well.

Bananas are on the list because they are known to cause migraines, and again, based on my clinical experience, cause problems for many people.

Finally, I advise cutting back on all meat and fish except veal and lamb. When yeast spoils meat, the toxic chemicals formed are worse than those formed by yeast in carbohydrates. In addition, chicken, turkey and pigs are fed cottonseed meal which is contaminated with a fungus called *Aspergillus.* I speculate that the animals store the *Aspergillus* poisons in their fat. This technique is a common way for animals to handle poisons. It is possible that storing the fungus poisons is one reason why yeast sensitive patients should not eat large amounts of meat. This is not a totally vegetarian diet, however, and we have many excellent recipes for cooking meat, fish and poultry in *Feast Without Yeast* and *Extraordinary Foods for the Everyday Kitchen.*

Removing wheat/gluten and milk/ casein: Stage III

This stage of the dietary changes takes you in a different direction. Because many people with yeast sensitivity also have gluten and casein sensitivity, I routinely advise patients who are still having problems to stop eating those foods.

What is gluten? Gluten (pronounced gloo'-ten) is a protein found in many grains, including wheat, rye, barley, oats, and some other grains. In the typical American diet, eliminating gluten means eliminating wheat. However, because so many other grains contain gluten, you will find that the main grain in your diet is rice. Corn is gluten free, but should not be eaten on a yeast free diet because corn is inherently mold contaminated. Other exotic gluten free grains include amaranth and quinoa. We do not use these grains and none of the recipes in *Feast Without Yeast* and *Extraordinary Foods for the Everyday Kitchen* use them. If you choose to use exotic gluten free

Allowable Foods on Stage III

Fresh and freshly frozen Meat, Fish and Poultry:
♦ all kinds (preferably hormone and antibiotic free). At this point, all types are permissible, but cut back on quantity.
♦ Canned Tuna Fish (labelled "very low sodium"--only ingredients Tuna Fish and Water)
♦ "Natural" hot dogs and luncheon meats only; nothing with sodium nitrates or sodium nitrites

Fresh Produce:
♦ all kinds except apples and apple products, grapes and grape products, mushrooms and bananas. These include all fruits and vegetables, fresh and dried herbs.

Dry Goods:
♦ Dried beans (all types)
♦ Tea
♦ Brown rice and all brown rice products
♦ Unprocessed clover honey

♦ Gluten free flour made from beans, rice or gluten free grains, such as amaranth and quinoa. No wheat, corn, rye, oat, barley flour.
♦ Rice cereals, such as puffed rice and cooked rice
♦ Potato chips that are not cooked in cottonseed, corn or peanut oil
♦ Oils: expeller pressed safflower oil, canola oil and olive oil
♦ Pasta (rice only)
♦ Sea salt
♦ Rice or potato based snack foods and other processed foods that are **free of**: wheat, milk, corn, rye, NutraSweet (aspartame); rice syrup, monosodium glutamate (MSG), malt, chocolate, yeast, vanilla extract, and vinegar

Butter and Eggs:
♦ Eggs, preferably hormone and antibiotic free
♦ Butter
♦ Ice Cream substitutes without chocolate or vanilla flavoring

grains, you need to be sure that you are not sensitive to them. You also should purchase the grains from a source that refrigerates the grain, particularly after grinding into flour, to be sure that the grain is not mold contaminated.

What is casein? Casein (pronounced ka'-sene) is a protein found in milk and dairy products. On Stage III, you would eliminate all dairy products except butter. These include: milk, cottage cheese, yogurt, and so on. Follow Stage III for four to six weeks, continuing to consult with your health care practitioner. Why remove casein and gluten? I explain the problems thoroughly in Chapter 4. Clinically, children with autism improve significantly when they remove casein and gluten from their diet. I will give a brief explanation here.

STAGE III LIST

REMOVE THE FOLLOWING FOODS FROM YOUR DIET:

All foods containing milk protein

> butter is acceptable except for extremely sensitive people—see your health care practitioner

All foods containing gluten, including wheat, oats, barley, and rye

Why remove Casein and Gluten?

Both casein and gluten contain substances that are toxic to the brains of sensitive people. When these substances are degraded by the body's digestive system, chemicals are released which resemble opioid chemicals. These chemicals are named opioids because they have effects on the brain similar to opiate drugs such as morphine, which affect pain systems. The brain has internal opioids which regulate the body's pain systems. The internal opioids are called endorphins. These food derived opioid chemicals are absorbed and react at receptors, or sites, for the body's own internal opioids.

The problem here is that opioids slow the brain down, so these chemicals hinder brain function. This is especially true for people with Autism, and may likely be true for people suffering from Attention Deficit Disorder and other problems. In addition, yeast grow well in dilute milk, so milk may aid yeast growth.

> **Using butter on a milk/casein free diet is controversial. Butter is the fat from the milk, so it should not contain casein. In reality, it might contain some very small percentage of milk protein. This does not affect most people. However, in cases of extreme casein sensitivity, you should not use butter. Substitute safflower oil for the butter. If you have questions about whether you or your child can use butter safely, consult with your health care practitioner.**

If you continue to feel that you need treatment, you probably are among the most severe in sensitivity. You then go on to Stage IV.

Very few patients go on to Stage IV--perhaps five to ten percent. That is only 5 or 10 people out of 100. However, we have designed *Feast Without Yeast* so it is safe and usable for everyone. About 80% of the recipes are suitable for people following Stage IV! *All* of the recipes in *Extraordinary Foods for the Everyday Kitchen* are gluten and casein free, as well as yeast free.

At this point, patients might consider testing or retesting their urine for yeast metabolites through Great Plains Laboratory, http://www.greatplainslaboratory.com.

> **You will need to examine labels even more carefully at this point. Many foods contain hidden casein.** For example, many brands of soy cheese and rice cheese are labelled "dairy free" because they contain no lactose (a milk sugar that commonly causes problems), but they contain casein. Some brands of canned tuna fish contain casein. Look at each packaged food carefully.

The Final Step: Stage IV

Stage IV is the final stage of this process of dietary change. It appears to be restrictive, but in reality you still can enjoy a wide variety of tastes and textures. More than 80 percent of the recipes in *Feast Without Yeast,* and all of the recipes in *Extraordinary Foods for the Everyday Kitchen,* are suitable for Stage IV. Look for recipes that are designated "Suitable through Stage IV" in the gray box under the recipe title.

On a personal note, our son, now a teenager, has followed Stage IV for several years. He is near the top of the growth charts in height and weight, and hardly ever gets sick.

STAGE IV LIST
REMOVE THE FOLLOWING FOODS FROM YOUR DIET:

Melons

Grapefruit

All meat except veal and lamb; occasional hormone and antibiotic free chicken and turkey are permissible

Yellow onions

leeks, scallions, garlic and spring onions are acceptable

Fruits except very fresh fruit in season, such as berries

Canned goods

Fish

Eggs, if necessary due to individual food sensitivity.

Many people find that they are sensitive or allergic to eggs.

Now that you are living a yeast free life, you should feel much more comfortable. Use your chart as a diary of how you or your child feel, behave, and otherwise live your life. If you make mistakes, look for how you can benefit from them.

Allowable Foods on Stage IV:

Fresh and freshly frozen Meat, Fish and Poultry:
♦ Veal and lamb are preferred. Occasionally, use hormone and antibiotic free chicken and turkey.

Fresh Produce:
♦ all fresh green vegetables except mushrooms
♦ lemons
♦ fresh and dried green herbs, such as basil, marjoram, dill, oregano, etc.
♦ raspberries
♦ blueberries
♦ blackberries
♦ cranberries
♦ boysenberries
♦ tomatoes
♦ potatoes
♦ sweet potatoes
♦ leeks, garlic, scallions and "spring onions"
♦ Mild chili peppers, such as Cubanel peppers
♦ Sweet (red) bell peppers

Dry Goods:
♦ Dried beans (all types)
♦ Brown rice and all brown rice products

♦ Unprocessed clover honey
♦ Gluten free flour made from beans or brown rice
♦ Rice cereals, such as puffed rice and cooked rice
♦ Oils: expeller pressed safflower oil, canola oil and olive oil
♦ Pasta (rice only)
♦ Sea salt
♦ Rice or potato based snack foods and other processed foods that are **free of:** wheat, milk, NutraSweet (aspartame); monosodium glutamate (MSG), malt, rice syrup, chocolate, yeast, vanilla extract, corn, rye and vinegar. For potato chips, examine each chip carefully to be sure it has no green edges or other signs of mold contamination.

Butter and Eggs:
♦ Eggs, if tolerated. Many people who require Stage IV diets also are sensitive or allergic to eggs. If you can eat eggs, try to find hormone and antibiotic free eggs.
♦ Butter, if OK with your health care practitioner

Notes

[1](Casal and Linares, 1981). Casal, M and Linares, M.J.. The comparison of six media for chlamydospore production by Candida albicans. *Mycopathologia.* 76(2):125-8, 1981.

[2]Law, B. A. Microorganisms and their enzymes in the maturation of cheeses. in *Progress in Industrial Microbiology,* ed. M. E. Buchell. Elsevier, Amsterdam v. 19, pp. 245-284, 1984.

Chapter 17

Nystatin

Nystatin is a nontoxic, anti-yeast medication that kills yeast as it travels through the gut. Nystatin is a natural substance, made by a bacteria that lives in the soil. The bacteria make nystatin to kill yeast in the soil. The name "nystatin" is a tribute to its discovery at a New York state laboratory. Nystatin has virtually no major side effects. It has been used for more than 40 years and is considered safe to use long term. I have patients who have taken nystatin for more than 10 years with no side effects.

Nystatin is not the same as the "one time only" anti-yeast pill, Diflucan, or other systemic antifungal medications like ketoconazole. Nystatin does not work in the bloodstream. Nystatin is the only anti-yeast medicine which goes all through the intestinal tract and reaches all the yeast. All the other anti-yeast drugs are absorbed high up in the intestinal tract and do not reach all the intestinal yeast. They are not as effective as nystatin is at killing the yeast. The other anti-yeast drugs also can have major side effects and cannot be taken long term.

After nystatin is taken, the Candida stops making its toxic chemicals and stops being a target for the body's immune system. Then the patient gets better and need not do anything else. Right?

No, it is not that simple. Nystatin does little without a special diet. As I discussed in Chapter 2, and as you have seen from Chapter 16, many foods contain toxic yeast chemicals which also kill bacteria. If a person continues to eat these chemical-laden foods, that person is taking in the equivalent of a low level of antibiotics all the time. Nystatin may kill some yeast, but the chemicals in the diet make room for the yeast by killing bacteria. Then the yeast comes back. And other foods in the diet help yeast grow better.

However, if you remove from your diet all the foods containing these chemicals, nystatin can kill the yeast. Then tremendous improvement can occur, as you have seen from all the cases in this book.

The best results from treatment come from eating the best anti-yeast diet, one which eliminates all the antibacterial chemicals and yeast growth factors which the yeast use either directly or indirectly to kill bacteria and make room for themselves to grow. The best anti-yeast diet is the 4 Stages diet. It is the easiest to follow, and is designed to take away the worst foods first. Nystatin works best by starting the 4 Stages diet at least a few days before starting the medication.

What about "die off"?

Die off refers to the problem that results when large amounts of yeast are first killed. An anti-yeast drug kills the yeast. Killing the yeast opens it up and releases all the toxic chemicals at once, so the person temporarily feels worse. This is often analogized to putting out a fire. When you put water on a fire, you get smoke. The smoke sometimes is worse than the fire, at least temporarily.

Patients who start with the 4 Stages diet and follow it for at least a few days before starting nystatin do not experience any significant die-off.

In my clinical experience, die off only occurs when a person takes nystatin without following an anti-yeast diet. I have not seen significant "die off" when patients follow my treatment plan. Why? If nystatin or any other antifungal drug is given without an anti-yeast diet, the yeast will be killed, but will then grow back because of anti-bacterial chemicals in the diet. When the yeast starts to grow again, it is very active. At these times, the yeast may release more toxic yeast chemicals. I believe it is this regrowth of yeast which people think is "die off." Fortunately, this regrowth does not occur when a patient follows my 4 Stages diet.

Nystatin is not absorbed and has no toxic side effects. If large amounts are taken at once, diarrhea may occur. At lower doses, however, this seldom occurs. I recommend a very gradual schedule of increasing doses of nystatin to prevent such problems.

Prescription information

Nystatin comes as a powder which may also be compounded into topical creams and gels, liquid suspensions and tablets. The pure powder is by far the best for successful treatment of yeast related disorders. I do not use the tablets because I am not sure if they break up well enough to do any good. The creams and gels can be useful for skin problems. I prescribe the liquid suspension to help children get started on nystatin, and occasionally for adults. Nystatin oral powder can be mixed into any liquid, dissolved in honey, or put into empty capsules and swallowed. It tastes bitter, so the sweeter the substance into which you put it, the better it will taste.

I recommend the following prescription.

> **R** *Nystatin oral powder, 150 million units. Take as directed on physician handout, Appendix C.*

This prescription allows you to purchase a bottle of nystatin powder which will last from four to six weeks initially depending on what dose you work up to. The dosing schedule for nystatin powder is given in Appendix C. The starting dose of the liquid suspension is 2 cc three times a day. I use this suspension to get children started on nystatin. Two and half milliliters (cc's) of the liquid equals 1/16 teaspoon of the pure powder. Children can take the liquid until they have transitioned over to the powder.

Duration of treatment

How long should you take nystatin? I usually advise taking nystatin for at least a few months after the symptoms are gone. In some cases, such as treating autistic children and patients with other difficult disorders such as multiple sclerosis, the patient may need to stay on nystatin long term. I have seen in those patients tremendous backsliding when nystatin is stopped. Because there are no harmful side effects from stopping nystatin, and because you can start nystatin easily, you can experiment with going off nystatin to see what happens. Sometimes you will see symptoms recur quickly, which will spur you on to get back on the program. However, sometimes symptoms will gradually recur. Since the backsliding in progress is not so dramatic, some parents who take their children off nystatin do not restart the nystatin. Sometimes all gains are lost. These cases, especially with autistic children, are the saddest because a small inconvenience of the anti-yeast diet could have allowed these children to live relatively normal lives.

When nystatin is stopped, yeast can come back and symptoms can recur. Recurrence is much less likely if one stays on the proper diet. If people feel uncomfortable or for whatever reason do not wish to continue taking nystatin, I recommend taking at least a few doses of one-eighth (1/8) teaspoon of nystatin powder per week. It is critically important before tapering off nystatin that one do this while in contact with your doctor. At a minimum, document exactly how you or the patient feels while on the full dose of nystatin. Then keep a record and compare each week to how you or the patient felt while still taking nystatin, to be sure you or the patient continue to feel as good. If there is any recurrence of symptoms, gradually increase the nystatin back to full dose, while making sure you or the patient follows the diet.

Using the correct dose

For most people, one fourth (1/4) teaspoon of nystatin powder four times a day is enough. A few people may need to take three eighths (3/8) teaspoon of the powder four times per day. Do not take nystatin without removing at least vinegar and malt. Do not start at too high a dose with too little dietary compliance. The nystatin dose must be balanced with dietary compliance to get the best results and prevent problems.

Clostridia

Dr. William Shaw, the clinical director of the Great Plains Laboratory, has looked at urinary levels of yeast chemicals in children treated with nystatin. He found that nystatin reduced these levels. But he also found that levels of chemicals from a different microorganism, the bacteria, Clostridia, actually increased with nystatin.[1] Clostridia is a microorganism, a bacteria, which grows well after other bacteria have been cleared out by antibiotics. Clostridia can also grow more if yeast is cleared out. Clostridia also make harmful chemicals. If a person takes too much nystatin and only follows the diet a little bit, then the person may not feel well. This may be due to Clostridia. For this reason, Dr. Shaw suggests keeping doses of nystatin on the lower side. He recommends a maximum dose of 2 milliliters three times per day. I agree that without the 4 Stages diet, Clostridia growth can occur above this dose. I note that Dr. Shaw's studies on yeast chemicals and nystatin were all done without a special diet.

If the patient does show Clostridia on a test, this can be treated without antibiotics. I recommend taking Lactobacillus GG for two to three weeks, continuing to follow the Four Stages diet, and continuing with a low dose of nystatin. Lactobacillus should be stopped after this two to three week period.

I have found that the 4 Stages diet prevents growth of Clostridia, so nystatin can be used at much higher doses than those recommended by Dr. Shaw.

How to take nystatin oral powder

Oral nystatin powder is just that, a powder. For adults or older children who can swallow pills, the patient or pharmacist can put the nystatin into empty capsules.

For the really hardy, nystatin can be dissolved in water and swallowed. Don't try this the first time you use nystatin. You'll never use it again. The taste is bitter. Keep in mind that mixing the bitter with the sweet makes everything bitter; mixing the bitter with water is a real experience. However, after you get used to nystatin, mixing nystatin with water is tolerable. I recommend this to people who can tolerate the taste. This approach clears yeast in the mouth and esophagus. So even if you take mostly capsules (see below), taking the powdered nystatin in water is helpful to do at least once per day.

The easiest way to take nystatin, for adults and people who can swallow pills, is to get empty gel capsules from your pharmacist or from a health food store. Twist open the capsule and fill the bottom part with nystatin powder. Then put the top back on. You can make several doses at a time and store them in your refrigerator.

Giving nystatin to young children

Young children cannot swallow pills and will not take nystatin in water. The easiest way to give nystatin to them is to mix it in unprocessed or "natural" honey. We have done this for many years. Mixing up several nystatin doses at a time takes about 10 minutes.

Directions for mixing up nystatin doses: Use empty, washed 35 mm film containers. These containers are great to use because they seal tightly. Place the appropriate dose of nystatin in the container. Take approximately 1/2 teaspoonful of unprocessed honey and drop it in the container. The honey should cover the nystatin, but not be too much. Using a chopstick, mix the nystatin and the honey. The nystatin dissolves in the honey. When it is time to give the dose, use a baby spoon to scoop out the nystatin. You can premix up to two days' worth of nystatin doses at a time. You can send the premixed doses to school, camp, or anywhere else.

For babies, start with the nystatin oral suspension. The dose is 2 ml (or 2 cc) 3 times per day.

Some young children do not want to take the nystatin mixed in honey. If you have that problem, use the nystatin oral suspension. Gradually start mixing the oral powder into it to increase the potency. Then gradually start shifting to nystatin mixed in honey.

My experience is that children associate taking nystatin with feeling better, especially when their parents help create that association. Most children don't have any problems taking nystatin. However, if their parents set the tone that this is "yucky" and "bad," children will have a hard time taking it.

Intravaginal use of nystatin

For chronic vaginal yeast infection or vaginal discharge or itch, use some #3 gelatin capsules, which you can purchase at your pharmacy. Fill a capsule half full of nystatin powder. If you cannot find #3 gelatin capsules, obtain whatever size is available and place one half of one eighth (1/2 of 1/8) teaspoon of nystatin in this gelatin capsule. Insert this capsule high up in your vagina twice a day for 1 to 2 weeks or for as long as necessary.

Other antifungal medications

There are several other antifungal drugs besides nystatin. These include ketoconazole (Nizoral) and fluconazole (Diflucan), as well as some newer ones.

These drugs do not work as well as nystatin for the kinds of problems I describe in this book. They are more useful for systemic candidiasis, such as for infections in cancer and AIDS patients. Even Diflucan, the "one pill" cure for vaginal yeast infections, does not, in my experience, work as well as nystatin for chronic and recurrent vaginal yeast infections.

Nystatin is not absorbed and is not toxic. Nystatin travels all through the gut and kills yeast lining the inner gut. The drugs ketoconazole (Nizoral) and fluconazole (Diflucan) are absorbed high up in the gut, then circulate in the blood. They cannot reach all the yeast which lines the gut. The fact that the drugs ketoconazole and fluconazole are absorbed high up and do not reach all the yeast makes it easy for the yeast to grow back. If the diet is not changed to exclude antibacterial chemicals, then this regrowth of yeast will be that much easier after the drugs ketoconazole and fluconazole are stopped.

The other big problem with ketoconazole and fluconazole is that because they are absorbed they have toxic side effects and really cannot be taken long term. Deaths have been reported from use of ketoconazole. Their use should only be temporary. Nystatin can be taken long term.

However, sometimes using ketoconazole is appropriate. For example, a major flare up of eczema or other skin rash due to dietary deviation may need to be treated with both an increase in nystatin and a few day's worth of ketoconazole. Another instance is when person has had a large exposure to some very bad chemicals, for example, when an autistic child eats plants covered with molds, or dirt. Usually a few days' worth of ketoconazole is sufficient to bring the child back to a tolerable level without suffering the medication's side effects.

Do not try to get ketoconazole (or nystatin) from a friend, however. These are prescription medications. You need to discuss all of these treatment decisions in consultation with your doctor. The main times one would consider using a few doses of ketoconazole are after antibiotic use, after steroid use, or after a child eats too much of the wrong food, for example, a whole tray of cookies at a party, a huge piece of malt-laden cake, or a lot of pickles.

Side effects of nystatin

As I stated before, nystatin has virtually no significant side effects. This is to be expected from a drug which is not absorbed. Nystatin may cause a few moments of nausea. Large amounts, taken suddenly, can cause diarrhea. There has been one case of a serious, rare rash reported from nystatin in the last forty years.[2] In reviewing this case, it is not clear that nystatin even caused this rash. Nonetheless, this side effect is noted in some drug reference books. Some books do not even mention this side effect, due to its rarity. [3] I note that many other kinds of drugs, such as antibiotics and anti-seizure drugs cause such rare rashes much more regularly. [4]

Nystatin cannot cause any direct worsening of symptoms because nystatin is not absorbed. However, when the body starts to clear the yeast, the body's adaptation to the yeast chemicals diminishes. People, including children, adapt over time to chronic pain and illness. When some pain is alleviated, other pain feels more acute. When a person starts to take nystatin, sometimes as the body clears the yeas, its chronic adaptation to the yeast and the pain yeast causes diminishes. Then foods which before were causing chronic problems, to which the person had adapted, now cause acute problems.

For example, an autistic child may have suffered chronic headaches. Removing headache causing foods relieves this pain, so now the child spends part or all of his time pain free. If this child now eats something which causes headaches, the child reacts acutely and may scream as if in terrible pain. A parent may believe the nystatin caused the screaming. In reality, the child now is in pain due to the onset of a new headache. Before the child was always in pain but had adapted to it.

To assess whether nystatin is the direct cause of this problem, you first need to look at the entire diet. Removing some problematic foods, but not all, causes the other foods to have stronger reactions. For example, if you remove vinegar and malt, but leave peanut butter, the child may feel much better until he or she eats the peanut butter. Then the child may be in agony compared with the time before they ate the peanut butter. The peanut butter is the problem, not the nystatin.

Second, you need to be sure you have not inadvertently increased the amount of something else to which the child is also sensitive. For example, a child going on the gluten/casein free diet may begin to eat a lot of corn, a gluten free grain. However, corn is frequently mold contaminated. Because the child also continues to eat other mold contaminated foods, however, one may not see any acute reaction. So now when the parent begins to eliminate the worst yeast offenders, the parent may see acute reactions to corn. The child's general chronic adaptation to feeling bad lifts when some yeast chemicals are removed, but when the child eats the corn, the effect is felt more severely. So the reaction is to other foods in the diet, not to the nystatin.

This does not always occur. It especially does not occur when starting the 4 Stages at the beginning.

The answer is not to stop nystatin. Instead one must look for the other offending foods and remove them.

What happens when treatment is stopped prematurely?

Let us go back to the case of Heidi, described in the first chapter of this book. Heidi came into my office with symptoms suggesting she might be developing a major developmental disorder, such as pervasive developmental disorder, an autistic disorder, characterized by children failing to interact appropriately. She started the treatment and signs of such a possible disorder disappeared. She was on the pathway to normal development, something that in my experience as a child psychiatrist never occurs spontaneously.

Heidi came back about six months later. Her parents had taken Heidi off the anti-yeast treatment. Heidi's symptoms had all returned. She now was not developing appropriately. Her parents had taken her to another doctor who had prescribed some standard psychiatric medications, dextroamphetamine, for her high activity. I happened to see Heidi because this doctor was away on vacation.

Heidi now had all the developmental problems I had feared when she had first come in. Heidi came in because she had a terrible reaction to the dextroamphetamine. I suggested to the parents that they restart the anti-yeast treatment because it had been so effective. They told me they did not believe that the anti-yeast treatment was responsible for the improvement they had seen. This was contrary to what they had told me in the office six months previously, as reflected in my notes. Nonetheless, they did not want to restart anti-yeast treatment, even though now their daughter was experiencing serious lifelong developmental problems.

Unfortunately, this case is not isolated. For reasons I cannot explain, some patients or parents of patients become ecstatic when they see major medical problems disappear. I had one patient, described in the book *Biological Treatments for Autism and PDD* (William Shaw, ed., Great Plains Laboratory), who had virtually lost her speech. She was down to two words. Within a week of starting nystatin, she was talking and singing. Yet six weeks after starting the nystatin, her parents decided to stop the medication. They never restarted it. The child lost all gains, and spent years in other therapies and special education, hoping to regain what they had in 6 weeks of nystatin.

Many parents stop treatment, the problems recur, and they then deny their initial observations. I could write a lot at this point about why I think this occurs, based on psychiatric training, but I won't. I will only say this: keeping a journal is critical. You as a patient or a parent need to be able to look at your own words to see what works and what doesn't. In my experience, your own words are the most convincing tool you have for keeping on track. If you stop taking nystatin as an experiment, first write down exactly how you feel. Look back at how you felt before you started the anti-yeast treatment. Keep checking to see how you feel and write down the feelings. If symptoms recur, start back on the nystatin.

The yeast Candida albicans can come back and cause the same problems it had caused before. Two of the women with multiple sclerosis described in Chapter 11, experienced complete resolution of symptoms. One experienced partial resolution of symptoms. All three stopped the treatment and their symptoms recurred.

Why does Candida come back?

Many foods contain malt which contains growth factors for yeast. Vinegar contains chemicals which kill bacteria and leave yeast alone. Many other foods are problematic, as you saw from the beginning of this chapter and in Chapter 2. Eating these foods enables yeast to come back. Other factors leading to increased yeast growth include taking antibiotics, taking steroids, eating other toxic substances such as dirt, or mold contaminated plants (a common problem in autistic children), and others.

How to prevent Candida from coming back

I recommend to patients that they stay on at least Stage I of the 4 Stages diet most of the time and take some nystatin. The dose of nystatin to prevent recurrence is lower than the dose necessary to clear out yeast in the first place. If one eats a meal containing the wrong foods, the yeast may come back a little, but getting back to the right food choices will prevent yeast from growing further. In other words, if there is a major family celebration such as a wedding, eat some of the food. Try to avoid the worst things, such as salad dressing and pickled foods that contain large amounts of vinegar. You may have a setback for a day, but one meal will not bring all the problems you had back. But if you find that eating a piece of birthday cake sets you back for weeks or months, your problem is very severe and you probably should avoid birthday cake altogether.

If you only are observing the proper food choices some of the time and if you start to feel your problems coming back, go back to the diet. If you are not taking nystatin, restart the nystatin. Restarting the treatment before your problems become bad again is much better than waiting. Although this sounds like a lot of work, it really isn't considering the alternatives. Most people agree that avoiding certain foods but living a normal life, is preferable to a life of pain and misery that most patients describe to me before starting treatment.

Tests for yeast

When I first started to do anti-yeast therapy, no reliable tests existed for the yeast Candida albicans. Now there is at least one. Dr. William Shaw of the Great Plains Laboratory measures yeast and fungal chemicals in the urine. Dr. Shaw was the laboratory director of a large children's hospital and was finding such chemicals in the urine of children with major developmental problems. Dr. Shaw now has his own laboratory. He measures a number of fungal chemicals in the urine as well as other chemicals reflecting different metabolic conditions. For yeast, his urine test is the best.[5] If you show a positive reaction, you know for sure you have a yeast overgrowth.

However, a lack of yeast does not mean no yeast exists. The test may have "false negatives." This means that a person showing no yeast chemicals in the urine may still have yeast. Dr. Shaw's test measures only some yeast chemicals, but not all. Not finding some of the chemicals does not mean for sure that no yeast is present. A change of diet and the use of nystatin may still be tried even with a negative urine test. This is especially true for autistic children. Your doctor needs to order the test from Great Plains Laboratory. You can check with your insurance company to see whether the test is covered.

Other companies test for yeast in the stool. I have not used such tests regularly and will not comment on them. Again, yeast may not be found on such a stool test, yet it could still be present, so a negative stool test does not rule out the presence of yeast.

Acidophilus

One product, Acidophilus, deserves more explanation. Many people believe that acidophilus provides good bacteria for the gut and can keep the yeast down. Doctors may tell you to eat yogurt while taking antibiotics, to recolonize the gut with "good" bacteria. You may get advice from nutritionists or other people that you do not need nystatin, just acidophilus.

Not only is this advice not true, it is actually bad for you after a few weeks.

Let me repeat. Acidophilus can be bad for you taken for more than a few weeks. I do not recommend it.

The research shows that acidophilus helps yeast grow. The research on the relationship between Candida and Lactobacillus, the main bacteria in acidophilus, shows that these two microorganisms like each other and grow quite well together. In fact there is research showing that Lactobacillus helps Candida grow.[6]

The reason that pure Lactobacillus GG seems to help temporarily is that Lactobacillus can crowd out another bad bacteria, namely Clostridia.[7] Lactobacillus does this within a few weeks. The lactobacillus treatment must be stopped after a few weeks. After this time, Acidophilus hurts rather than helps.

I have had many patients continue taking acidophilus while on the anti-yeast diet and nystatin. For awhile, they got better, but then their improvement stopped or reversed. Once they stopped acidophilus and usually other supplements, the improvement restarted.

Anti-yeast supplements and Megavitamins

Many people prefer taking supplements to taking nystatin. Among those who prescribe supplements are people who are unable to prescribe nystatin, because they are not medical doctors. Other people simply do not like taking any prescription medication.

There are at least two types of supplements to consider when looking at Candida. The first type consists of all the various "natural" products sold as yeast fighters, including grapefruit seed extract, garlic, oregano oil, etc. I am not an expert in the use of these products. I know that nystatin is much stronger and works much better than any of these products at killing yeast, so I only prescribe nystatin. I cannot recommend any of these other products.

The second set of supplements is used more often in autistic children. These are megadoses of magnesium and vitamin B-6. These megavitamins have been helpful to many children and adults. The question is, why? And do you need to continue them while on an anti-yeast diet?

The vitamins may be helpful in part because they lower yeast chemicals. One of the yeast chemicals, spermine, is toxic to bacteria and is also toxic to humans.[8] This chemical replaces magnesium in important cellular structures.[9] Moreover, vitamin B-6 binds with this chemical in such a way as to neutralize it.[10] It is possible that reduction of spermine is one reason why magnesium and vitamin B-6 help sometimes. The problem is that there are many other toxic yeast chemicals besides spermine. In my clinical experience, treating for yeast reduces all the toxic yeast chemicals. This is far better than trying to neutralize one toxic yeast chemical. What I find is that the effects of anti-yeast treatment outshines the effects of megavitamin treatment, **and** megavitamin treatment may interfere with anti-yeast treatment.

Currently, I recommend that no vitamins be used when starting anti-yeast therapy. I have seen a number of children who were taking both high doses of vitamins and nystatin. In most of these children, nystatin seems not to work. I suspect vitamin B-6 or some other supplement may bind to nystatin, making it inactive. These children improved after stopping the vitamins. There is no reason to use high doses of vitamins together with nystatin and anti-yeast therapy.

Another major problem with using vitamin and mineral supplements is they feed the yeast. The megavitamins may make yeast much harder to treat. I now tell parents and patients that if they wish to continue megavitamins that they cannot assume nystatin will work. Autistic children who take both megavitamins and nystatin have a very uneven course with some things getting better, some things getting worse, and usually very little overall improvement. On the 4 Stages diet and nystatin only, autistic children consistently improve.

Conclusion

In conclusion, I recommend starting the 4 Stages diet, explained in Chapter 16, beginning with Stage I-A. Continue slowly and gradually, taking careful notes as you do this. By the end of Stage I, you should be on nystatin. Follow the instructions in this chapter and Appendix C for taking nystatin!

If you want to experiment with going off nystatin, be sure that you write down exactly how you feel. Then you will be able to know with confidence whether your symptoms return. If they do, just start back on the diet and nystatin. You'll feel better again in a short time.

Notes

[1]Dr. Shaw published a book on his lab test and his findings in *Biological Treatments for Autism and PDD*, Second Edition, 2002, published by the Great Plains Laboratory, 11813 W 77th St., Lenexa, KS 66214. The phone number is 913-341-8949.

[2]Stevens-Johnson Syndrome Associated with Nystatin Treatment, Arch Dermatol. 1991 May; 127(5):741-2.

[3]The major manufacturers of nystatin, including Lederle and Paddock, do not even include Stevens-Johnson as a possible adverse reaction. See package inserts for those products. The American Hospital Formularly Service states that nystatin has rarely caused hypersensitivity reactions. Nystatin is generally considered a non-toxic and safe medication with few side effects. American Hospital Formulary Service Drug Information Book, 2001.

[4]Stevens-Johnson syndrome, although rare, is much more often associated with antibiotics including amoxicillin and erythromycin. Limauro, D.L., Chan-Tompkins, N.H., Carter, R.W., Brodmerkel, G.J., Jr., Agrawal, R.M., "Amoxicillin/ clavulanate-associated hepatic failure with progression to Stevens-Johnson Syndrome;" *Ann Pharmacother* 1999 May; 335):560-4; Anderson, J.A.,

"Antibiotic drug allergy in children," *Curr Opin Pediatr* 1994 Dec; 6(6):656-60; Chan, H.L., Stern, R.S., Arndt, K.A., Langlois, J., Jick, S.S., Jick, H., Walker, A.M., "The incidence of erythema multiforme, Stevens-Johnson syndrome, and toxic epidermal necrolysis. A population-based study with particular reference to reactions caused by drugs among outpatiens," *Arch Dermatol* 1990 jan; 126(1):43-7; Shoji, A., Someda, Y., Hamada, T., "Stevens-Johnson syndrome due to minocycline therapy," *Arch Dermatol* 1987 Jan; 123(1): 18-20; Pandha, H.S., Dunn, P.J., "Stevens-Johnson syndrome associated with erythromycin therapy," *NZ Med J* 1995 Jan 25; 108 (992):13; Gorbachev. V.V., Bronovets, I.N., Vetokhin, V.I., Chugunkina, S.B., Karachan, N.Z., "Stevens-Johnson syndrome as a complication of antibacterial therapy," *Ter Arkh* 1972 Aug; 44(8):106-8; Curley, R.K., Verbov, J.L, "Stevens-Johnson syndrome due to tetracyclines-a case report (doxycycline) and review of the literature," *Clin Exp Dermatol* 1987 Mar; 12(2):124-5; Crosby, S.S., Murray, K.M., Marvine, J.A., Heimbach, D.M., Tartaglione, T.A., "Management of Stevens-Johnson syndrome," *Clin Pharm* 1986 Aug; 5(8):682-9.

Stevens-Johnson syndrome has also been associated with anti-epileptic drugs. Pelekanos, J., Camfield, P., Camfield, C., Gordon, K., "Allergic rash due to antiepileptic drugs:clinical features and management," *Epilepsia* 1991Jul-Aug; 32(4) 554-9; Bocquet, H., Farmer, M., Bressieu, J.M., Barzegar, C., Jullien, M.< Soto, B., Roujeau, J.C., Revuz, J., "Lyell syndrome and Stevens-Johnson syndrome cause by lamotrigine," *Ann Dermatol Venereol* 1999 Jan; 126(1) 46-8; Chan, H.L., Stern, R.S., Arndt, K.A., Langlois, J., Jick, S.S., Jick, H., Walker, A.M., "The incidence of erythema multiforme, Stevens-Johnson syndrome, and toxic epidermal necroylsis. A population-based study with particular reference to reactions caused by drugs among outpatients," *Arch Dermatol* 1990 Jan; 126(1):43-7; Crosby, S.S., Murray, K.M., Marvine, J.A., Heimbach, D.M., Tartaglione, T.A., "Management of Stevens-Johnson syndrome," *Clin Pharm* 1986 Aug; 5(8):682-9.

[5] For more information about the Great Plains Laboratory, please call 913-341-8949 or check the website at www.greatplainslaboratory.com.

6 Isenberg, H. D., Pisano, M. A., Carito, S. L., and J. I. Berkman. Factors Leading to Overt Monilial Disease. I. Preliminary Studies of the Ecological Relationship Between *Candida albicans* and Intestinal Bacteria. *Antibiotics and Chemotherapy.* 10(6): 353-363, 1960.

7 Biller, J.A., Katz, A.J., Flores, A.F., Buie, T.M., and S.L. Gorbach. Treatment of Recurrent *Clostridium difficile* Colitis with *Lactobacillus* GG. *Journal of Pediatric Gastroenterology and Nutrition.* 21:224-226, 1995.

8Stevens, L. Regulation of the biosynthesis of putrescine, spermidine and spermine in fungi. *Medical Biology*, 59, 308-313, 1981.

9Weiss, R. L., and D. R. Morris. *Biochimica Biophysica Acta.* 204:502511, 1970.

10Keniston, R. C. Cabellon, S. Jr., and K. S. Yarbrough. Pyridoxal 5'Phosphate as an Antidote for Cyanide, Spermine, Gentamicin, and Dopamine Toxicity: An in Vivo Rat Study. *Toxicology and Applied Pharmacology.* 88:433-441, 1987.

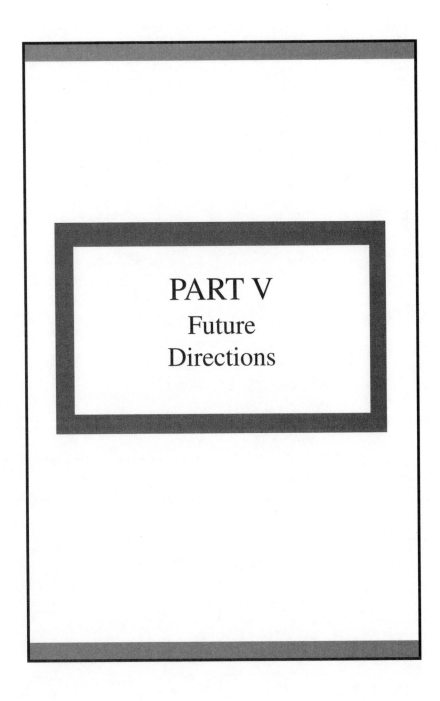

PART V
Future
Directions

Chapter 18

Taking Charge of Your Health

I have treated many patients for Candida albicans, and have heard from many more who followed my program using our book, *Feast Without Yeast*. [1] The ones who follow the program benefit. Patients have been cured of supposedly incurable conditions such as multiple sclerosis. Patients with autism, debilitating fibromyalgia and arthritis, among others, have improved so significantly that they are able to lead relatively normal lives.

These are patients who have taken charge of their health, either personally or through their parents and caregivers. They have sought out and followed the anti-yeast treatment I describe here. They have overcome obstacles, including people who tell them they can't possibly feel better because this treatment is different from prescribed standard treatments. They have bucked the standard medical system. Many of these patients have continued to follow anti-yeast treatment with their own doctors. I congratulate them!

I have treated many other patients long term. I congratulate them, also!

All of these people are feeling better because they took charge of their health.

Unfortunately, I also hear from many of my patients who started taking charge of their health, but were discouraged by the skepticism of their family, friends and doctors. Some people became so discouraged that they quit treatment, and went back to their misery.

Beware the Skeptics and Cynics!

Patients tell me that despite feeling better than they have felt in months or years, their regular doctors or family and friends simply cannot believe that changing diet and conquering yeast can be so effective. These skeptics and cynics ask, "If this treatment is so great, why doesn't my doctor (or why don't I) know about it? If my doctor (or I) doesn't know about it, it must not be so great. So if it isn't so great, it doesn't really work." The skeptics and cynics then tell these patients, "Your improved health is all in your head," or "you really weren't sick to begin with," or "your condition just naturally improved."

Of course, there is no logic at all to these thoughts. Good doctors, and good friends, know that they should evaluate a treatment based on what it does, not based on what they have or have not heard.   They will have an open mind about new treatments.   One sign of a good doctor is his or her ability to listen to a patient who has found a new treatment that works. This doctor will investigate and draw their own conclusion about that treatment, not just demean and demoralize a patient for taking charge of their own health. Many doctors are excellent in this respect.

I do not mean to imply, in this chapter, that your particular doctor is close minded, and I congratulate you if you have found a great, open minded doctor. However, I wrote this chapter because I hear all too often about the problems my patients have with medical professionals, family and friends who are not open minded.

Encountering skeptics and cynics is unfortunate not just for the patient's emotional well being, but for their health. Many patients adopt these thoughts. They go off the 4 Stages diet, and they feel sick once again. Then for reasons I cannot fathom, they conclude that the treatment did not work.

One patient came to see me to treat numbness in her hands. This is a serious condition, a possible precursor to multiple sclerosis. This patient worked in a large medical center affiliated with a medical school. After following my treatment, the numbness disappeared. Treatment was effective. This patient enthusiastically told her colleagues about this. Unfortunately, her colleagues were not excited about her improved health. They told her how upset they were that she had "sold out" by seeing someone outside mainstream medicine. Her supposed offense against mainstream medicine, which had done nothing for her health condition, was so great that this patient's colleagues did not even care that she was better. Fortunately, this patient did not go off the 4 Stages diet and nystatin and she continued to improve.

The saddest cases I see responding to skepticism are those involving children. Children cannot choose their own medical treatment; their parents choose the treatment and can choose to abandon the treatment. I have seen far too many cases where this happens. I remind you of the case of Heidi, described in Chapter 1, and whom you also visited in Chapter 16. Heidi was a three year old girl who was developing autistic symptoms. She did extremely well with treatment. Her parents stopped the treatment. Less than a year later, she came in with diagnosable developmental delays, a characteristic of autism. When I suggested that they restart the anti-yeast treatment, they responded that it had not worked, despite the clinical notes taken at the time showing that the treatment had worked, and the notes showing how happy they were about the treatment. Sadly, this child could have lived a normal life.

I had another case where the diet and nystatin had restored a child's speech. I will call her Barbara. Barbara had autism, and had rapidly been losing her speech. When I saw her, she was down to two words. Barbara was singing and talking by the end of the first week of following the 4 Stages diet and taking nystatin. After about six weeks, Barbara's parents decided to "experiment" by taking her off the nystatin and discontinuing the diet. Barbara's progress immediately began to slow. Eventually she lost most of her gains. Barbara's parents did not restart nystatin because Barbara, age 3, did not want to take it easily. They convinced themselves that she really did not make the gains she had made. Barbara, several years later, still struggles with the same problems. Barbara's parents had allowed a three year old to make major medical decisions that will affect her the rest of her life.

Skepticism is rooted in our cultural outlook on disease

Why are some people so skeptical about this treatment?

I believe the skepticism stems from our basic cultural outlook on healing. In our western, American culture, we learn that the key to health is killing disease. We either cut it out through surgery or take drugs. The second best tactic is to suppress the symptoms by taking drugs until we think we feel better.

When medical conditions do not respond to drugs or surgery, though, as is the case with most illnesses that respond to anti-yeast treatment, traditional medical training does not provide much for the average doctor to fall back on. We doctors do not learn much in medical school about therapies that are alternatives to traditional Western medicine. For many doctors and patients, accepting the fact that drugs or surgery are not the only answer may be difficult.

There are many reasons for this.

Medical training and aggressive drug company marketing encourages doctors to look at drugs as the only way to conquer disease

First, as I said above, we live in a drug oriented culture, and doctors are part of that culture. Doctors spend huge amounts of their medical training learning how to use standard drugs. The pharmaceutical industry invests heavily in marketing these drugs to doctors. Every day, drug "reps" visit countless medical offices to discuss their products privately with doctors. They bring along slick charts showing how effective drugs can be. This sell job has gone on for years. Drug companies give out "freebies," ranging from pens, pencils and calendars with their product name emblazoned on them, to books, dinners, and entertainment. They send out free "educational" materials, including books, pamphlets, videos and CD's about the problems their product treats, implicitly encouraging the doctor to prescribe the product. They give out soccer balls for the doctors' children, emblazoned with the name of the drug and the drug company. All of this is in the name of education.

Drug companies now market drugs directly to the public. You can hardly watch TV or open a magazine without seeing ads for the latest, most expensive drugs to treat asthma, allergies, depression, anxiety disorders, and other common problems, urging you to "ask your doctor" about this, that or the other prescription drug.

Why do they do this? Because new drugs are very profitable. The pharmaceutical industry invests in researching and developing drugs, and has a certain number of years during which the medication is on patent (meaning that there is no generic competition). The industry must maximize its profit from those drugs during the time on patent. The drug company sales representatives usually do not push drugs that have gone off patent, because those drugs no longer are very profitable. They face competition from generic, or non-name-brand equivalents.

Drug companies don't aggressively market nystatin

What is to prevent a doctor from looking into other types of treatment, such as anti-yeast treatment? Doctors have invested years in learning how to use medications and the marketing forces of the drug companies are strong. The drug companies that sell nystatin are not sending out reps, pens, pencils and fancy information. They do not sponsor dinners and cruises. They do not send out tapes, CD's and videos in the mail to educate doctors about yeast problems, and encourage prescribing nystatin. Nystatin has long been off patent. Nobody goes around touting to doctors the benefits of having their patients eliminate vinegar and malt from their diets.

Medical specialization discourages looking at groups of disorders as having a common cause and common treatment

The second reason for doctors' skepticism lies in the specialized nature of medical treatment. Medical training and practice is geared toward treating very specific illnesses with very specific treatments performed by superspecialists. For example, neurologists see patients who suffer from migraines. Dermatologists see patients who suffer from skin problems. Doctors, and the medical industry, are not used to thinking that medical conditions that may have different types of symptoms, such as eczema and migraine headaches, may actually have a unifying cause, yeast, and a unified treatment, changing diet and treating the yeast. This is more often than not the case in patients I see.

Patients rarely come to me with only one problem, as you have seen from the cases presented. In treating a single problem, their other problems resolve. So treating yeast challenges the narrow, specialized focus of modern medicine.

So the drug oriented medical culture, the narrow nature of medical thinking, as well as the pharmaceutical industry, can make doctors skeptical about any treatment that is out of the mainstream, including changing diet.

Doctors dislike the lack of organized research on treating yeast, despite the high benefits and virtually nonexistent risks of anti-yeast treatment

The third reason for doctors' skepticism involves the state of research, or lack of research, about treating yeast. Many doctors will say that not enough is known about the yeast problem, so they won't treat it and won't believe that this treatment can help their patients. They correctly point out that nobody has done any large-scale controlled, double blind tests of changing diet and using nystatin. Without those studies, they say, they can't support such a treatment.

I, too, would love to have controlled, double-blind studies to refer to. While in theory one should always have this gold-standard research to back up treatments, this is not always possible. I discuss the reasons why in Chapter 19 and below in this chapter. It is helpful to remember that many long accepted treatments have been used without the benefit of these double blind tests. For example, for a long time, nobody knew how aspirin worked. Yet we used it for pain relief because it was effective. Many major surgeries were not subjected to double blind studies. For example, cleaning out cholesterol laden plaques from the arteries going up to the brain for years was not subject to an efficacy study.

You should also understand that large, broad, double blind studies are not necessary for all treatments, and are not necessary before beginning to treat yeast. The reason is that the risks of treating yeast are very minimal, and the benefits potentially are enormous.

Changing diet has no medical risk

The risks of changing diet are about as low as you can get. The only real harm in changing diet is slight inconvenience. Changing diet is cheap. You need to eat; you may as well eat good foods. The benefits of changing diet are high, as I describe in this book. If you feel that changing diet doesn't work for you, then you have only lost a few weeks of eating some foods that you may like, and can switch back at any time.

Taking nystatin has almost no medical risk

The risk in taking nystatin, as I explain in Chapter 17, also is almost zero. The only real side effects of nystatin are, for the occasional patient, nausea or gastric upset, and this dissipates within a few days. Some people see the benefits of this treatment almost immediately. In more involved cases, patients usually see benefits within a few weeks. So each patient can evaluate the treatment on their own. Again, if the patient believes that the treatment isn't worth the work, they can stop taking nystatin immediately with no ill effects.

Controlled studies may never be done; meanwhile, you feel worse and worse when you could feel better and better

The problem with waiting for controlled studies is that you may wait forever. Meanwhile, you will continue to feel terrible. The incentives in the research industry all are against this research ever being done. Research costs money. Money comes from private sources, or the pharmaceutical industry, or the government. Since most research doctors cannot fund their own research, they rely on drug companies or the government to fund research.

To get this funding, however, doctors need to fit their research into what those companies, or the government, are interested in funding. The trend in research over the last few decades has been devoted to understanding body processes on a cellular and genetic level so that we can find targets for drug therapy. The research funding comes from government agencies that are very traditional in choosing research projects, and drug companies.

Diet—what you eat—does not fit this mold of current research trends.

The fact that there are no large scale studies showing this treatment is effective should not stop you from trying this treatment, and should not stop your doctor from supporting it.

No valid studies disprove anti-yeast treatment!

Why should the lack of studies *proving* anti-yeast treatment works not stop you from trying it? Because there also are no valid studies *disproving* this treatment. You're on neutral ground. In addition, there is plenty of other evidence, in the way of individual case reports, that the treatment works. That is why I wrote this book.

Most doctors have not been trained to understand diet, so they are not accustomed to thinking that the foods we eat affect our health.

The fourth reason for skepticism among doctors is that they hold the same belief as the majority of the public: that chemicals in the diet are unlikely to have much impact on health. Except for a few special illnesses for which dietary change now is routinely prescribed (for example, diabetes and high cholesterol), doctors usually are not trained to think that a patient's diet affects their health.

One very sad, but common, example of this occurred in my own family. I have a relative who developed terrible stomach problems when he was around a year old. He also started to get asthma. He was vomiting several times a day. He was about to start going to specialists, getting expensive and uncomfortable tests, and asthma medication, when his pediatrician, to his credit, suggested as a possibility taking this person off cow's milk. Guess what? The problems all disappeared almost immediately. What would have been treated as a life long, involved and expensive medical condition was actually an allergic reaction to cow's milk. How many children suffer unnecessary treatments because their doctors were not so insightful?

Even with the dramatic improvement in this relative's health, some specialists still did not believe it and wanted to subject him to more tests! They could not believe that food could have such an impact on health.

Yet I see this every day of my medical practice.

There is little institutional support for doctors recommending alternative or complementary treatments in general, not just anti-yeast treatment

The final possible cause for your doctor's skepticism is lack of institutional support for recommending a therapy outside the medical mainstream, commonly called "alternative" or "complementary." You need to understand the basis for that lack of support. It starts in medical school.

Medical training is grueling. Doctors endure long years of training that consists of memorization, long shifts, many times more than 24 hours at a time, tests, and daily oral quizzing in front of other doctors to make sure that he or she knows the state of modern medicine. Doctors also must take comprehensive examinations called "boards" during medical school, after medical school, and to be certified and recertified as specialists. The boards all test in-depth knowledge of standard medical procedures.

When I was in medical school, even raising questions about alternative medicine was considered such a breach of training that the question could put a student at risk of being kicked out of school.

The bias against "alternative" treatments may have relaxed somewhat as doctors have become increasingly aware that their patiens are using "alternative" treatments, and want to know what their patients are doing. More continuing medical education is devoted to these alternative and complementary treatments. However, I hear from many of my patients that their doctors still do not accept the type of treatment described in this book and their doctors are not willing to support the patients in this treatment.

Your doctor may perceive any request to use a therapy that he or she did not learn about in medical school as threatening to his or her fundamental training, which is the basis of his or her identity, as well as livelihood.

Doctors don't want to get sued

In addition, your doctor may perceive that any therapy outside the community standard of medicine puts him or her at risk of a malpractice suit. If you sue your doctor, your doctor could get kicked off of managed care and insurance lists. All of this is going through your doctor's mind when you ask for help with a yeast problem.

Anti-yeast treatment is safe, legal, and easy to monitor

Should such concerns be going through your doctor's mind? The main medicine for yeast is nystatin, a non-absorbed medicine which has almost no risks. Nystatin is perfectly legal. You are free to change your diet on your own. For a malpractice suit to be successful, there must be some harm. The risk of damage from nystatin is minute, and you should know about that in advance. There can be no damage from changing diet.

I also provide, in Appendix A, a sample Informed Consent form that you should go over with your doctor and sign, if you are comfortable with that. This form says that you understand the possible risks and benefits of treatment, and protects both you and your doctor.

Your doctor can overcome these systemic barriers because he or she wants you to get well and there are no other good options for healing you

This picture looks pretty discouraging. I have outlined numerous, systemic reasons, why doctors and other people may be skeptical about this treatment and may try to discourage you from trying it. You are asking your doctor to consider a set of problems in a way that he or she did not learn about in medical training. Your doctor learned that microorganisms such as bacteria, yeast and fungus, can cause considerable disease, but your doctor did not learn this particular pattern of illness and he or she did not learn this use of nystatin. He or she cannot find good clear studies in medical journals. His or her colleagues do not use this method of treatment. Your doctor may fear getting sued. This lack of a learned pattern can be enough for the doctor to turn you down.

Despite all this talk about skepticism, a lot works in your favor with your doctor. Almost all doctors wish to help their patients. Sometimes they may not seem to care, but really they do. Despite all of the focus in medical school on specific treatments and therapies, doctors learn that their main goal is to heal. Doctors want to do something for you.

Second, for most of the conditions I describe in this book, doctors do not have other good traditional medical options. Most of the conditions I see, and that I discuss in this book, have not responded, and do not respond, to traditional medical treatment. Doctors are much more willing to try an unconventional treatment when no conventional treatment works.

Third, standard drugs cause enormous side effects. If a doctor wishes to treat depression with most of the usual drugs, for example, the drug may treat the depression but may also cause bad side effects, including sexual problems. Many patients are non-compliant with their medications due to these side effects. Anti-yeast therapy has virtually no side effects and it helps with sexuality at the same time it treats problems such as depression.

My book will help your doctor

I wrote this book and provided my cases and explanations to fill your doctor's gap in knowledge and to provide the theoretical basis for this treatment. You can give your doctor a copy of the book, or just the pages that apply to your medical condition. Have your doctor place it in your chart, if that would help you. I hope that such explanations will help your doctor see that such anti-yeast therapy has a basis.

You also should give your doctor a copy of the informed consent form that I supply in this book, in Appendix A. "Informed consent" means that you have been informed of the risks of the treatment, and you are consenting to that treatment. All treatments need informed consent before you can start them. This is why before starting a treatment, your doctor always tells you about the risks and benefits of the treatment. Occasionally, a doctor will do this in writing to make sure you understand the treatment you are undertaking, and to protect the doctor from a malpractice suit.

The Informed Consent form in Appendix A was substantially approved as part of two clinical research protocols I wrote, and which were approved by a medical school, but which unfortunately were not funded. I thought that such a form might be useful to help a patient talk with his or her doctor about obtaining a prescription for nystatin. If you acknowledge in writing to your doctor what you expect the risks and benefits to be of being treated for yeast, that you acknowledge that there may be other treatments out there, and that you wish to go with a change of diet and nystatin, your doctor will feel more secure about prescribing the treatment. You are taking responsibility for doing something outside the mainstream and your doctor is legally covered with your signature on this informed consent form.

All of this information should help support your doctor's decision to treat you in the best possible way

You have a special relationship with your doctor, and my purpose is to support that relationship so that you can take care of your health. I have explained why doctors have not been trained about the treatment I describe, and may be skeptical. However, I also hope that you understand that your doctor wants to help you. If you approach your doctor with knowledge and with the informed consent document to allay your doctor's fears, chances are good that he or she will help you.

If your doctor is not interested, go to a different doctor who can help you

If your doctor is unable to help you, you could choose to start treatment with another doctor, like myself. I understand that switching doctors is difficult, but your health comes first. Sometimes, people come to me for a consultation, then go back to their primary doctor to continue treatment. The primary doctor sees the benefits of treatment, so he or she supports it.

If even at this point your doctor is unwilling to support you, and you feel that you are benefitting from this treatment, you could think about finding a different doctor.

You are the patient; you are suffering; take charge of your health!

Always remember, you are the patient and it is your health, not your doctor's. You need to take charge of your health. You are not a controlled experiment. You are the one who has suffered. You are the person who is feeling better. You are a person who deserves to live the best, healthiest life possible. Remember that it is very easy for other people to give advice about your health when the person giving advice does not experience your problems. If you have the support of your doctor and family and friends, that is wonderful. If you don't, but feel better, continue the treatment. Because your job is to take care of yourself. I hope that this book helps you do that!

Notes

1 Semon, B. A. and L. S. Kornblum. *Feast Without Yeast: 4 Stages to Better Health.* Wisconsin Institute of Nutrition, 1999; 1-877-332-7899; http://www.nutritioninstitute.com.

Chapter 19

The State of Research on Yeast--where do we go from here?

To understand the yeast Candida albicans, we need to know about how it interacts with the human body. We know that many people have excess yeast in their bodies, and that these people have medical disorders that can be treated with anti-yeast treatment. We know from the research of Dr. William Shaw that some yeast chemicals can be found in the urine, and that these chemicals decrease after anti-yeast treatment. The symptoms of the disorders also decrease after anti-yeast treatment.[1]

We need to know more about the chemicals yeast makes and releases. We need to know the toxic effects of these chemicals. We need to know about yeast's interactions with the body's immune system. We also need to know about all the ways people introduce yeast into our bodies through the food we eat and the medications we take. I described in Chapter 8 how antibiotics clear space out for the yeast. I described in Chapter 2 how chemicals in certain foods clear out space for the yeast. Our diet contains enough anti-bacterial chemicals to make space for yeast in the intestinal tract. We also know that some people are reluctant to take action to protect their health unless scientists know everything about the problem and how to solve it.

What I have done in this book is present in an understandable form what we know about yeast, and my observations, based on the studies and my clinical experience with many patients, about what kinds of problems yeast cause and how we can solve those problems.

At this point we can ask, how much do scientists know about these topics?

Unfortunately, not enough.

I view as the largest area of need for research finding out more about the chemicals yeast makes, how those chemicals relate to our diet, and the long-term toxicities of those chemicals. Although we can identify many of the chemicals which yeast make, we do not know their long term toxicities. We particularly do not know how these chemicals add up and interact to be poisonous to the developing brain in fetuses and children and adults. We do not know if these toxicities are worse because these chemicals come in groups, and may interact with each other to cause increased toxicities. This lack of research is unfortunate. These chemicals may play large roles in causing headaches, autism, attention deficit hyperactivity disorder and many other disorders and in helping the yeast to stay in the intestinal tract.

Doctors like myself who treat patients with yeast problems know that clinically, many health problems respond to anti-yeast treatment. The treatment has virtually no side effects, and potentially has tremendous benefits. I have described those benefits in this book.

The issue I address in this chapter is the broader social issue about what information has been published about Candida in scientific journals, and why we are unlikely to get more information, given the current research climate.

The research on Candida: basic facts

Scientists used to research the basic science of Candida more vigorously. A book was published in 1964 collecting all the research on Candida albicans.[2] My favorite study from this book is the study on growing tuberculosis and Candida albicans together. Researchers studied a simple question: do they grow better together and help each other or not? The answer is that they grow better together. The capsule on the outside of Candida albicans is actually a growth factor for tuberculosis.

This research finding is extremely important. Candida albicans can help other "bad bugs" grow. Yet this finding was not pursued vigorously and certainly was not taught in medical school.

What happened to the research on Candida?

What happened to such research? Why did research on Candida slow down? In the 1950's only a few antibiotics were available. Now many antibiotics are available. More important, antibiotics are profitable to the drug companies that make them. If doctors stopped prescribing antibiotics, or slowed down their rate of prescriptions, because of justified fears about Candida infections, companies would lose money. This ties in directly with how research is funded, as I discussed in Chapter 18.

This is more than just a question of money, though. The lack of research on Candida reflects the nature of how research is done in this country.

Researching yeast requires looking at the "big picture," unlikely to be funded

In the future, more research on Candida may occur, but the amount will most likely be limited. Why? One reason is a simple one. Candida is hard to research and understand.

The amount of Candida in the body and in the intestinal tract is hard to measure. The chemicals it produces are hard to test for. The interactions of Candida with the body's immune system are complex.

Anyone interested in researching the health problems which Candida albicans causes would need to know about many diverse disciplines, including chemistry and immunology. The big picture of yeast is much larger than one study on one aspect of yeast. In other words, studying Candida requires researchers to study the "big picture" and cross disciplines. These factors make yeast less attractive to researchers to study. Let me explain further.

Few research studies cross disciplines. I know of no immunology studies on yeast which even considered the diet or the chemicals in the diet. Yet such interactions are exceedingly important in treating Candida effectively.

Researchers are unlikely to be able to study this big picture. Research is organized along disciplinary lines. People who study chemistry do not study physiology. People who study immunology stay with immunology and do not generally study nutrition or biochemistry. This is how research is organized and funded. The more specific and narrow the question, the easier time the scientist has funding for that research.

Researchers are connected with universities. Researchers are pressured to get grant funding to maintain their positions, and to publish papers. A "big picture" subject like how Candida affects health is unlikely to attract funding. The questions are too large, the studies would take too long, and the whole idea that yeast and diet could cause health problems is too foreign to modern medicine to attract grant funding.

Research about yeast chemicals

As I stated above, I think that one of the most important areas for concern is how the yeast chemicals interact and cause problems for us. So let us look specifically at research on yeast chemicals.

I have discussed in Chapter 2 how Candida makes a number of toxic chemicals, including phenyl ethyl alcohol. As I discussed in Chapter 2 (and see the references in that chapter), yeasts in general are known to produce a whole variety of toxic chemicals including toxic alcohols, aldehydes, acetates and the powerful brain poison hydrogen sulfide. This information comes from research studies on "off flavoring agents" of alcoholic beverages. Yeast make chemicals which are then found in alcoholic beverages. The brewery and winery people would prefer not to find such chemicals, because they make alcoholic beverages taste bad. These chemicals also happen to be toxic, and most of them are much more toxic than ethyl alcohol itself. So industry has figured out how to control the bad flavor of the yeast chemicals. They first had to study the chemicals to do so.

Studies on yeast chemicals made during production of alcoholic beverages constitute the majority of studies on the chemicals yeast produce. One would hope that someone would also look at these yeast chemicals in the context of studying alcoholism. A whole government agency studies alcoholism. However, the agency appears to study only ethyl alcohol, not other just mentioned toxic yeast chemicals. The off flavoring agents and other alcohols, actually are much more toxic than ethyl alcohol. Should government sponsor research on some of these other chemicals? Of course. Alcoholism is a major problem in this country. We should know everything we can about what alcoholics take in.

Why does nobody study these other chemicals?

Again, we need to look at funding. The original studies about yeast chemicals were funded by donations from the alcoholic beverage industry. Their purpose was to figure out how to make wine and beer taste better. These companies had no mandate to study toxicology or the effects of the chemicals on human health. They did not receive money to understand the effects of yeast chemicals on the brain. Now that the questions about off-flavoring agents have been answered, the original donors have no reason to fund more studies.

We need clinical studies of anti-yeast treatment

In addition to the need to study yeast chemicals, other aspects of yeast need study. The main problem here is the lack of broad-based clinical studies of anti-yeast treatment, to devise the best treatments for people who have problems which yeast cause, and hopefully to prevent further problems in those people and in others.

For reasons I do not understand, the medical community behaves as if this research were complete. Instead of looking at clinical benefit versus clinical risk of a proposed treatment, many health care providers will look only at whether a "double blind" study exists to support an alternative treatment. This viewpoint assumes that any important promising therapy will be researched in this country and researched with all the resources that the new therapy deserves. They reason, if there was anything to diagnosis and treatment of Candida albicans, studies would show it. What these doctors do not recognize is that, as I will discuss below, this assumption is false. The viewpoint that anything worthwhile naturally would have been studied is unrealistic and has no basis in reality. The research community has no automatic system for making sure that promising new therapies get the attention they deserve. there has been no good study *disproving* anti-yeast therapy as effective.

The net result is there have been no large scale studies of the effects of anti-yeast the treatment, period.

Drug companies are unlikely to fund anti-yeast studies

Drug companies research new drugs because they profit from new drugs. The major medicine for yeast, nystatin, no longer has patent protection. It is a relatively cheap drug, and widely used for specific yeast problems. Companies making nystatin do not make large profits from this drug. These narrow profit margins do not usually allow companies to fund large expensive studies.

Government is unlikely to fund anti-yeast studies

Theoretically, government could fund such studies of yeast, but it has not. Why not? The nature of such research funding precludes such research.

Only university affiliated researchers can apply for federal grants. A university hiring committee will look at the grant getting potential of any new hire. The hiring committee knows that there has to be interest in any new idea a potential hire is thinking about. The result is that someone with a new but unproven idea does not stand a chance of getting hired to work in a university. The hiring committee will not take someone who has no chance of getting grants.

Any person interested in research these days knows this. Grants are scare and getting more so. The only way to stay in research is to attach oneself to someone and a research area that is already being done. While following this approach may help one get hired at a university, this system also guarantees that research will stay with the status quo. What is being done is likely to continue being done. Even established researchers must show that their programs are generating data to continue getting grants. The best way to do that is to stay with what works and not take risks.

All grant applications for federal money go through committees called study sections. These study sections are made up of researchers from the same areas as the grant writer. These study sections collectively assign priority scores to each grant application. This procedure means that people in the field of inquiry as represented by the people on the study section would have to be interested in the new therapy for the grant to get a high priority score. Convincing one person will not do. There must be a collective interest. Second, study sections like to see preliminary data that shows that the grant writer has already done some of the work and that this therapy will yield useful data. So someone has to have some initial interest to spare up some money to try something new at least a little.

Established researchers are judged by the number of grants they get, which in turn depends on how many papers the researcher is able to publish. Researchers who publish more papers are assumed to be better than researchers who publish fewer papers. Thus there is little incentive for an established researcher to take a risk on something new which might not work and for which the researcher may not be able to publish a paper. So even if a research trainee comes to an established researcher and says "I would like to try something new," the answer is almost always no because of the risk of not being able to publish a paper. New ideas can get stopped at many places along the way.

What about smaller funding agencies? Smaller funding agencies, even so called alternative funding agencies, frequently put on their boards the same researchers as are at universities. The result is the same priority in funding.

All of these problems confront a researcher potentially interested in Candida albicans and the benefits of treating Candida albicans. The result is very little research on the potential benefits of treating Candida. The only real hope is private funding.

When your doctor tells you that this "yeast thing" has been researched and it does not work, he or she has no basis to make such a statement.

A cited study does NOT prove that nystatin does not work!

There is, however, one study about Candida, unfortunately often cited, to "prove" that Candida is not a problem. This study was published in the *New England Journal of Medicine* in 1990.[3] Reading the study, however, shows it really did not prove anything.

This study was poorly designed. The implied study question was, is nystatin effective in helping certain women feel better? This is how the study has been interpreted. Unfortunately, the actual study question was totally different. This question was, did certain women feel better *while* they were taking nystatin? The study design was flawed and did not allow for an honest answer to the real question, of whether nystatin helped these women feel better. The best way to have studied this question would be to evaluate women before they took nystatin, then give them nystatin to see if it helped, then determine whether a clear difference emerged from before and after nystatin. This study did not do that.

The *New England Journal* study was not designed to answer even its own question. To be sure, the investigators evaluated women both when they were taking nystatin and when they were not taking nystatin. However, the investigators failed to take into account the fact that some women were treated with nystatin first and others second. When the investigators questioned the women who supposedly were "off" nystatin, they did not distinguish at the time of questioning between the women who had taken nystatin and were currently off it, and the women who had never taken nystatin. So women who were "off" nystatin could have been treated with nystatin already and could have benefitted from that treatment. In other words, the study was flawed because there was no genuine control group. Some women who were considered to be "off" nystatin had already been treated with nystatin. As we know from the cases discussed in this book, treatment with nystatin kills the yeast and helps people feel better. So it is not surprising that the study did not detect significant differences between the two groups, because really there were not two genuinely different groups.

In addition, the study was flawed because the researchers did not use any type of anti-yeast diet. I (and others) have found that an anti-yeast diet is essential to making nystatin effective. The investigators recognized that they might have found more significant results if a diet had been used. However, these investigators did not want to use a diet because this would have made their study too complex, so diet was not included as a variable.

Notwithstanding their own reservations about the lack of an anti-yeast diet, the investigators made sweeping generalizations that their negative result (meaning no difference between the supposed control group and the treated group) provides additional "objective evidence" that the Candida hypersensitivity syndrome was not a "verifiable condition." In other words, they generalized from their poorly designed study that Candida is not a real problem, and that nystatin is not a real treatment.

This study did not prove anything. Their study provided no evidence whatsoever about whether Candida does or does not cause problems for people.

Normally we would hope that such a study would not be accepted by such a prestigious journal. The design was terrible, and no effect was shown.

Why was this study published? To find out, we need to look at an editorial published several months later, also in the *New England Journal*, which cheered the result.[4] The editorial writers were happy this study was negative.

What this shows is that Candida is as much or more of a political issue than a purely medical issue in the medical and research world. Acknowledging that Candida might be a problem would force doctors to reevaluate all the antibiotics they use. This also would require doctors to look at what we eat--diet--as important in their patients' health. Instead, the medical establishment clings to the belief that someone must have looked at this Candida thing, when in reality, no one in the academic medical world has studied it in a systematic, well-designed manner.

In summary, sufficient research does not exist on the yeast Candida albicans. No one can legitimately say that there are studies which show that Candida does not cause health problems.

Conclusion

What we know about the yeast Candida albicans, combined with the clinical results of treating for this yeast, strongly support anti-yeast treatment that I describe in this book. This is not the end of the story, however. Much more research should be done about how yeast affects our health. This research would help us prevent the kinds of devastating health problems many people suffer. The risks of treatment are almost nonexistent; the potential benefits are high. And, most important, no study has shown that Candida is *not* a health risk.

Notes

[1] *Biological Treatments for Autism and PDD*, 2002, second edition by Dr. William Shaw. The Great Plains Laboratory. For more information about testing for yeast, please contact Great Plains Laboratory, http://www.greatplainslaboratory.com.

[2] Winner, H. I. and Rosalinde Hurley, *Candida albicans*. J and A Churchill, London, 1964.

[3] Dismukes, W.E., Wade, J.S., Lee, J. Y., Dockery, B.K., and J.D. Hain. "A randomized double blind trial of nystatin therapy for Candidiasis hypersensitivity syndrome." *New England Journal of Medicine*. 323(25):1717-23, 1990.

[4] "A controlled trial of nystatin for the Candidiasis hypersensitivity syndrome." *New England Journal of Medicine*. 324(22):1592-94, 1991.

Chapter 20

Conclusion

I know of no more powerful therapy than anti-yeast therapy for problems ranging from chronic fatigue syndrome to multiple sclerosis than anti-yeast therapy. I am always amazed when I see how well it works and I hope that I never stop being amazed. The fact that the treatment of Candida albicans is not accepted is one of the truly great tragedies in American medicine and for society as a whole.

Let us be optimistic. If all women of child bearing age were treated for yeast and if children were treated for Candida during and after receiving antibiotics, we would not be seeing the epidemics of autism and ADD/ADHD that we now see. If everyone avoided foods that increase yeast, and if yeast were treated after antibiotics were given, then disorders such as multiple sclerosis, ulcerative colitis, psoriasis and all of the other disorders I describe in this book, would become rare. If those disorders were treated right away by anti-yeast therapy, people would not have to suffer with them for their entire lives.

I hope that this book is one step in this direction.

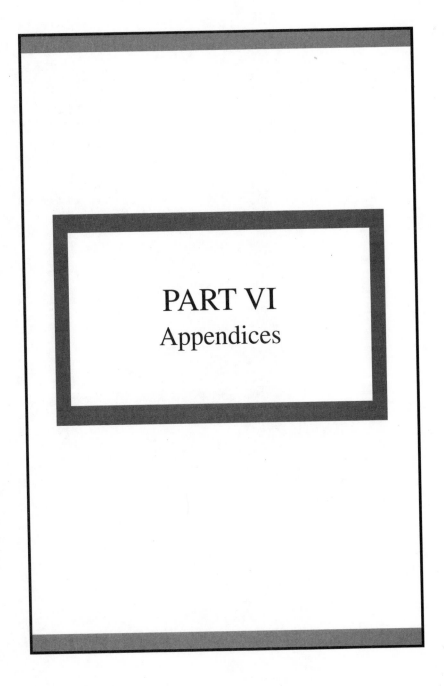

PART VI
Appendices

Appendix A:
Informed Consent Form for Prescription of Nystatin
Copyright 2003, Bruce Semon, M.D., Ph.D. and
Wisconsin Institute of Nutrition, LLP
1-877-332-8977
http://www.nutritioninstitute.com

The following is a sample Informed Consent form that was developed and approved for a research protocol. This is not intended to create specific legal advice for any particular person, nor is it intended to create an attorney/client privilege. Please consult your attorney if you have particular questions.

I would like to take nystatin for treatment of the following medical condition(s):
_____ I understand that nystatin is not the standard medical treatment for this condition. I understand that there are no academic studies to support the use of nystatin in such a condition. During my treatment I acknowledge that I will restrict my diet to exclude highly fermented foods such as vinegar, malt, alcoholic beverages, chocolate, pickles, aged cheese, soy sauce, miso, tempeh, worcestershire and cottonseed oil, and will follow the dietary principles outlines in *Feast Without Yeast: 4 Stages to Better Health*, and *An Extraordinary Power to Heal*, by Dr. Bruce Semon and Lori Kornblum.

Risks: I have been informed of the discomforts and risks which I may reasonably expect as part of this treatment. These include that I will not be able to eat all foods I may wish to eat. The diet will not be compromising in any nutritional way. Nystatin may cause some mild and transitory gastrointestinal distress including nausea, vomiting and diarrhea. These effects are infrequent. Nystatin is not absorbed into the body by the intestine. There is one reported case study of a possible severe reaction to nystatin, called Stevens-Johnson syndrome, which is a reaction also associated more commonly with many other drugs. However, the risk of this reaction is so low that the major manufacturers of nystatin do not even consider it a possible risk. If I begin to experience any problems, I will contact my doctor immediately.

Nystatin and dietary restriction of this kind poses no known risk to a fetus.

I acknowledge that treatment for my condition listed above consists of medications such as (circle the most appropriate treatments) prednisone, steroid cremes, antibiotics, antidepressants and other psychiatric medications, other changes in diet, other headache medicines, nonsteroidal antiinflammatory drugs, antihistamines, surgical procedures, and other (please specify).

_____

_____

_____

 I acknowledge that there may be other diagnostic tests which my doctor has recommended which I may not yet have obtained. I acknowledge that my doctor has explained these treatments and tests to me and that I have decided instead or in addition to use nystatin and anti-yeast diet treatment. I understand that using the 4 Stages diet and nystatin does not preclude further testing regarding my disorder.

The rationale for the use of the 4 Stages diet and nystatin for my condition described above is explained in the book *An Extraordinary Power to Heal*, by Dr. Bruce Semon (2003: Wisconsin Institute of Nutrition). I have read the relevant pages of the book. The book, or copies of the relevant pages, may be made part of my medical record.

Dated this _____ day of _____, 20____.

_____   _____
Patient   Physician

Date	Current behavior, behavioral changes, health and health concerns	Planned changes (what foods we're going to eliminate, other changes, etc.)	Any other changes in routine, medications, etc.

Appendix C:
Dosing Schedule for Nystatin Oral Powder
copyright 2003 Wisconsin Institute of Nutrition, and
Bruce Semon, M.D., Ph.D.

The information in this Appendix is intended for general use and may not apply to particular people. You may need different doses than recommended below. Each person should consult their doctor before taking any medication and before discontinuing any medication.

Notes: An eighth teaspoon (1/8 tsp.) of Nystatin is about 500,000 units. When "twice in the day" is recommended, that means spaced evenly—e.g., take one at breakfast and one at dinner. Three times a day might be breakfast, lunch and dinner. I recommend taking nystatin after you have eaten something, to avoid possible nausea.

Week 1 -

Day 1   1/16 teaspoon once in the day

Day 2   1/16 teaspoon twice in the day

Day 3   1/16 teaspoon three times in the day

Day 4   1/16 teaspoon four times in the day

Days 5, 6 and 7: 1/16 teaspoon four times in the day

Week 2 -

Day 1   1/8 teaspoon once in the day, 1/16 three times

Day 2   1/8 teaspoon twice in the day, 1/16 two times

Day 3   1/8 teaspoon three times in the day, 1/16 1 time

Day 4   1/8 teaspoon four times in the day

Days 5, 6 and 7: 1/8 teaspoon four times in the day

***Week 3 (First alternative)-

Day 1   1/4 teaspoon once in the day, 1/8 three times

Day 2 1/4 teaspoon twice in the day, 1/8 two times

Day 3 1/4 teaspoon three times in the day, 1/8 one time

Day 4 1/4 teaspoon four times in the day

Days 5, 6 and 7: 1/4 teaspoon four times in the day

****Week 3 (Second alternative)-

Day 1 1/8 tsp. 5 times per day

Day 2 1/8 tsp. 6 times per day

Day 3 1/8 tsp. 7 times per day

Day 4 1/8 tsp. 8 times per day

Days 5,6,7: continue at 1/8 tsp. 8 times per day

****Two alternatives are listed for week 3 because many people have a hard time taking 1/4 tsp. at a time. Smaller amounts are easier to take. Stay on the dose of week 3 or the dose which works well for you. The conversion from liquid nystatin to powdered nystatin is 2.5 cc liquid = 1/16 tsp powder. The usual dose of liquid to start is 2 cc 3 times per day.

Copyright information: Please feel free to use this information about Nystatin Dosing for personal use, and to copy and distribute it. Appendix C on Nystatin Dosing may be reprinted as an entire page on any other website or for conferences, as long as the website or conference properly credits the author. If reprinted on a webpage, the reprint must contain a link back to http://www.nutritioninstitute.com. This page may only be copied for other publications with the author's express permission. Please contact us at the above numbers for permission.

Appendix D: Travel and Restaurant Tips for Yeast Free/Wheat Free/Milk Free Diets

copyright 2003, Wisconsin Institute of Nutrition
1-877-332-7899
http://www.nutritioninstitute.com

1. Carry food with you for the entire trip.

2. Double the amount of food you would expect to eat on an airplane. Expect delays!

3. If the airline serves food, order special meals. The Diabetic meal usually is safest, with very plain food, no dressings or sauces.

4. *Hotel Survival*:

--Stay in a hotel with a kitchen

--Find corporate apartments that are fully furnished.

--Cook food in advance and stay in hotel with refrigerator and microwave.

--Last Resort: pack a single electric burner, electric frying pan and slow cooker. Follow fire regulations. Cook outside the hotel. Keep food in a cooler with ice. This works for 2 to 3 days.

5. *Restaurant enjoyment*:

--Check the menu in advance bu calling the chef and discussing your special food concerns, where possible.

--Order in advance or off menu: plain baked potatoes, brown rice with nothing added, salad without dressing (fresh lemon wedges), fresh vegetables, etc.

--Ask the chef for ingredients, stressing you have food allergies.

Copyright information: Please feel free to use this information about Travel and Restaurant Tips for personal use, and to copy and distribute it. Appendix D on Travel and Restaurant Tips may be reprinted as an entire page on any other website or for conferences, as long as the website or conference properly credits the author. If reprinted on a webpage, the reprint must contain a link back to http:// www.nutritioninstitute.com. This page may only be copied for other publications with the author's express permission. Please contact us at 1-877-332-7899 or email us at support@nutritioninstitute.com for permission.

Appendix E: IEP Tips for Children on Yeast Free/Wheat Free/Milk Free Diets

copyright 2003, Lori S. Kornblum, Attorney at Law and Wisconsin Institute of Nutrition
1-877-332-7899
http://www.nutritioninstitute.com

This information was created by Lori S. Kornblum, an attorney in Milwaukee, Wisconsin. We hope that this is helpful information for children with special needs who are on special diets. This information is not intended to be specific legal advice, because we do not know you or your child, nor is it intended to create an attorney/client relationship. If you need particular legal assistance, please contact your attorney.

An IEP is an Individualized Education Plan, for education of children who qualify for special education under federal law. If your child has an IEP and is following a specialized diet, we recommend that the IEP contain information that makes the entire school aware of the dietary restrictions. In the IEP, designate in the appropriate place that the child is on a special diet for his or her condition that necessitates the IEP. For example, if the child has Autism, specify that the child is on a special diet for treating his or her autism. State that specific information is attached, then YOU write the attachment.

OPTION A [the safest strategy]
"[Child] is on a special diet to treat the following condition:_____. [Child] may eat only the following foods:

1. Child may eat only foods that are sent from home.

2. If other foods are available in school as treats or snacks, child may eat the following: [list all of the foods the child CAN eat]

3. Staff may not supplement the child's diet in any way, as treats, incentives or for any other reasons, without first obtaining parental consent, unless the food is listed above.

OPTION B [risky because you are dependent on people being as vigilant as you are]

[Child] is on a special diet to treat the following condition: _____[Child] may not eat any foods containing the following ingredients: [list those foods]

OPTION C: [risky for the same reason as Option B]

[Child] is following Stage [I, II, III, IV] of *Feast Without Yeast: 4 Stages to Better Health* and *An Extraordinary Power to Heal.* Attached are the pages of the book that explain the diet. School must follow this plan.

After choosing Option A, B, or C, all IEP's should contain this information:

--Staff shall inform parent as much as possible about upcoming parties, treats, school projects, etc., involving food, so parent can send in appropriate treats for [child].

--[Child] may not share or trade food. Staff must supervise child during snack and meals to ensure that child eats safely. Any deviation from [child's] diet will likely result in severe behavioral symptoms, which may include:

--Kicking, hitting, biting, head banging, pinching, scratching, grabbing, stomach pain, head pain, lack of cooperation, hyperactivity, lack of concentration [list any other that you have experienced].

--Staff shall be trained and made aware of the dietary restrictions and the child's diet. If [child] eats something that he/she is not supposed to eat, which results in one or more of the behavioral problems listed above, the school, will not suspend or expel the child, because such behavioral problems are a direct manifestation of [child's] condition of _____.

Copyright information: Please feel free to use this information about IEP's for personal use, and to copy and distribute it. Appendix E on IEP information may be reprinted as an entire page on any other website or for conferences, as long as the website or conference properly credits the author. If reprinted on a webpage, the reprint must contain a link back to http://www.nutritioninstitute.com. This page may only be copied for other publications with the author's express permission. Please contact us at 1-877-332-7899 or email us at support@nutritioninstitute.com for permission.

Appendix F: Shopping list for a Yeast-Gluten-Casein Free Diet

Staple foods to keep in the house at all times:
____Beans (garbanzo, pinto, lentils, kidney, lima and navy)
____Rice (long and short grain brown rice, red rice)
____Rice flour
____Whole wheat flour and whole wheat pastry flour, if not on Stage III
____Expeller-pressed safflower oil
____Garlic (keep in freezer)
____Elephant garlic (keep in freezer)
____Sea salt
____Fresh ginger
____Dried herbs: basil, oregano, dill, marjoram, sage, rosemary, and other green herbs
____Rice pasta

Shopping list:

Vegetables:
____tomatoes (all kinds, and lots of them)
____carrots (baby and long for cooking)
____lettuce (romaine and red, or two other kinds)
____broccoli
____green beans
____zucchini
____cubanel or other mild pepper

____red bell peppers, if in season
____seasonal vegetables, such as asparagus, cauliflower, green beans, brussel sprouts, summer squash, etc.
____parsnips
____leeks
____scallions
____other fresh onions in season
____one red onion
____potatoes (red and russet)
____seasonal potatoes (Yukon gold, superior, other varieties)
Fruit in season
____pears
____oranges
____berries except strawberries
____lemons
Dry goods:
____rice flour
____whole wheat flour (not on Stage III or IV)
____oatmeal (not on Stage III or IV)
____rice pasta
Dairy
____butter
____eggs
____rice milk
____goat cheese, if tolerated
____goat's milk, if tolerated
Extras
____**Treats for members of the family not following 4 Stages**

copyright 2003, Wisconsin Institute of Nutrition 1-877-332-7899, http://www.nutritioninstitute.com. Personal use permitted and encouraged. This list may be reprinted on any website or for conferences. Reprint must credit authors and link back to http://www.nutritioninstitute.com. This page may only be copied for other publications with the author's express permission.

Index

Quick Order Form

Mail Orders: Wisconsin Institute of Nutrition, LLP, 5555 No. Port
Washington Road,Suite 200, Glendale, WI 53217

Telephone Credit Card Orders: Toll-Free 1-877-332-7899; local or
long distance: (414) 351-1194

Title	Price	No. Books	Total
An Extraordinary Power to Heal	$24.95	_____	$_____
Extraordinary Foods for the			
Everyday Kitchen	$15.95	_____	$_____
Feast Without Yeast	$ 22.95	_____	$_____

Sales Tax (Wisconsin Residents) 5.6% $_____

Shipping and Handling: $5.50 for the first book

plus $3.50 for each additional book to the same

address: $_____

Outside of US **or** when ordering more than three of any one title, please call
for ordering information and discounts.

                                        TOTAL AMOUNT DUE $_____

PAYMENT: _____ Check enclosed   _____VISA _____MASTERCARD

Name on Credit Card: _____

Card Number: _____ Exp. date: ___/____/____

ORDERED BY AND SHIP TO (PLEASE USE ONE FORM PER AD-
DRESS, OR WRITE THE INFORMATION ON THE BACK):

Name: _____

Address: _____

City: _____ State: _____ Zip: _____

e-mail: _____

phone number: (_____) _____